# The Stability of Dynamical Systems

J. P. LASALLE
Brown University

# Appendix A
# Limiting Equations and Stability of Nonautonomous Ordinary Differential Equations

Z. ARTSTEIN
Weizmann Institute of Science

**SOCIETY for INDUSTRIAL and APPLIED MATHEMATICS**

PHILADELPHIA, PENNSYLVANIA 19103

*Printed for the Society for Industrial and Applied Mathematics by*
*J. W. Arrowsmith Ltd., Bristol 3, England*

# Contents

**Chapter 4**
ABSTRACT DISCRETE DYNAMICAL SYSTEMS AND PROCESSES. NONAUTONOMOUS DIFFERENCE EQUATIONS



**Appendix A**
LIMITING EQUATIONS AND STABILITY OF NONAUTONOMOUS ORDINARY DIFFERENTIAL EQUATIONS
***by Zvi Artstein***

# Preface

To some extent it is true that in the western world Liapunov's Direct Method was rediscovered in the mid-1950's. At least by that time its importance in the design of nonlinear control systems had been widely recognized. My understanding and appreciation of Liapunov's theory began in 1959 when Solomon Lefschetz and I wrote an elementary text on the subject. It was in the process of writing that book that I discovered a simple relationship between Liapunov functions and Birkhoff limit sets. This observation provided a unity to Liapunov's theory and greatly extended his direct method. This has had many ramifications beyond ordinary differential equations and has been the subject of much research during the past decade. The purpose of these lectures is to present an introduction to these newer developments.

We begin in Chapter 1 with the simplest of dynamical systems—the discrete semidynamical systems associated with autonomous difference equations—and we see in this elementary context the main ideas and structure of the general theory. In Chapter 2 we carry out the development of the analogous theory for autonomous ordinary differential equations (local dynamical systems). Chapter 3 is a brief account of the theory for retarded functional differential equations (local semidynamical systems). Here the state space is infinite-dimensional and not locally compact. Among the most recent developments has been the discovery of invariance properties of the limit sets of the solutions of a broad class of nonautonomous ordinary difference equations. A discussion of these invariance properties and their relationship to stability theory is given in Appendix A written by Zvi Artstein. Chapter 4 is a presentation of this same theory for nonautonomous difference equations. This, while new, is naturally more elementary.

I would like to thank, first of all, John R. Graef who organized this regional conference at Mississippi State University and who, by inviting me, provided the motivation for preparing these lectures. I enjoyed the well-run conference and the enthusiasm of the participants. I am grateful to Zvi Artstein for the lectures he gave there and for allowing me to include his notes as an appendix to this volume.

J. P. LASALLE

*Little Compton, Rhode Island*
*January* 1976

# CHAPTER 1

## Difference Equations. Discrete Semidynamical Systems

**1. Introduction.** Today there is more and more reason for studying difference equations systematically. They are in their own right important mathematical models. Yet very little is required other than an understanding of convergence and continuity, and there are no troublesome questions concerning the existence and domain of definition of solutions. Moreover, their study provides a good introduction to the stability theory of differential equations, difference–differential equations, and functional differential equations. We shall see in this chapter, not only all of the basic features of the general theory, but also some results that are new. A good introduction to the classical Liapunov theory of stability for difference equations with applications to control system design and analysis is [49] (see also [81], [82], [32]). In [46] Hurt goes beyond classical theory and gives applications to numerical analysis. We shall extend Hurt's results. In Chapter 4 we consider nonautonomous difference equations.

**1.0. Notation.**

$J$ is the set of all integers.

$J_+$ is the set of all nonnegative integers.

$R^m$ is $m$-dimensional Euclidean space with $\|x\|$ the Euclidean norm.

We allow the usual, and convenient, notational ambiguity that $x$ may denote either a vector or a function. Let $x: J_+ \to R^m$. Then $x'$ and $\dot{x}$, functions on $J_+$ to $R^m$, are defined by

$$x'(n) = x(n+1),$$
$$\dot{x} = x' - x,$$
$$T: R^m \to R^m.$$

The difference equation

(1.1) $$x' = Tx$$

stands for

$$x(n+1) = T(x(n)) \quad \text{all} \quad n \in J_+.$$

The solution to the initial value problem

(1.2) $$x' = Tx, \qquad x(0) = x^0$$

is $x(n)=T^n(x^0)$, where $T^n$ is the $n$th-iterate of $T: T^{n+1}=T(T^n)$ and $T^0=I$, the identity mapping. The product of functions is composition. Equation (1.2) is simply an algorithm defining a function $x$.

**1.1.** *Exercise.* Show that the $m$th-order difference equation $(g: R^m \to R)$

$$u(n+1)=g(u(n), u(n-1), \cdots, u(n-m+1))$$

is equivalent to a system of $m$ first order difference equations (1.2).

**2. Discrete dynamical systems on $R^m$.** We shall assume from this point on that $T$ is continuous.

**2.1.** DEFINITION. A *discrete dynamical system* on $R^m$ is a mapping $\pi$: $J_+ \times R^m \to R^m$ satisfying for all, $n, k \in J_+$ and all $x \in R^m$:

(i) $\pi(0, x)=x$,
(ii) $\pi(n, \pi(k, x))=\pi(n+k, x)$,
(iii) $\pi$ is continuous.

Every difference equation (1.1) defines a dynamical system $\pi: \pi(n, x^0)=T^n x^0$, and, conversely, every discrete dynamical system has associated with it the difference equation $x'=Tx$, where $T(x)=\pi(1, x)$. For this reason we confine our attention to the difference equation (1.1). The *motion* $T^n x$ *from* $x$ refers to the sequence of states $x, Tx, \cdots, T^n x, \cdots$. Condition (ii) is the semigroup property and expresses the uniqueness of the solution in the forward direction of time. $\pi$ is often called a "semiflow" or a "semidynamical" system, and the term "dynamical system" is used when $J_+$ can be replaced by $J$ ($T$ has an inverse).

**3. Limit sets of motions.** There are many reasons for being interested in what happens to $T^n x$ for large values of $n$. This concern with the asymptotic behavior of $T^n x$ is what stability theory is all about. More than that, stability has to do with dynamic behavior. What happens under both perturbations of $x$ and $T$? Liapunov's definitions of stability have to do only with perturbations in $x$. Later in § 11 we shall show how this relates to perturbations in $T$.

Basic to methods of successive approximation of solutions of $x=Tx$ is the fact that, if $T^n x^0$ converges, its limit is a solution (a *fixed* or *invariant point*). What we do now is to generalize this fact. We first introduce the notion, following Birkhoff [15], of the limit set $\Omega(x)$ of $T^n x$. (We are interested only in positive $n$ and drop the adjective "positive.") Then in the next section we shall show for bounded $T^n x$ that $\Omega(x)$ is invariant $(T(\Omega(x))=\Omega(x))$. In § 6, and in what follows, we shall show how Liapunov functions can be used to obtain information about the limit sets of motions.

**3.0.** NOTATION. For $x \in R^m$ and $S$ any set in $R^m$

$$\rho(x, S)=\inf\{\|y-x\|; y \in S\}, \tag{3.1}$$

the *distance of x from S*;

$T^n x \to S$ as $n \to \infty$ means $\rho(T^n x, S) \to 0$ as $n \to \infty$;

$\bar{S} = \{x; \rho(x, S) = 0\}$ is the *closure* of $S$.

A set $S$ is *closed* if $\bar{S} = S$ and *open* if its complement is closed.

$T(S) = \{T(x); x \in S\}$.

**3.1.** DEFINITION (Birkhoff). A point $y$ is a *limit point* of $T^n x$ if there is a sequence of integers $n_i$ such that $T^{n_i} x \to y$ and $n_i \to \infty$ as $i \to \infty$. The *limit set* $\Omega(x)$ of the motion $T^n x$ from $x$ is the set of all limit points of $T^n x$.

**3.2.** *Exercise.* Show that an alternate definition for $\Omega(x)$ is

$$\Omega(x) = \bigcap_{j=0}^{\infty} \overline{\bigcup_{n=j}^{\infty} T^n x}. \tag{3.2}$$

**3.3.** *Exercise.* For any $H$ in $R^m$ define

$$\Omega(H) = \bigcap_{j=0}^{\infty} \overline{\bigcup_{n=j}^{\infty} T^n(H)}. \tag{3.3}$$

Show that $y \in \Omega(H)$ if and only if there are sequences $n_j \in J_+$ and $y_j \in H$ such that $T^{n_j} y_j \to y$ and $n_j \to \infty$ as $j \to \infty$.

## 4. Invariance.

**4.1.** DEFINITION. Relative to (1.1), or to $T$, a set $H$ is said to be *positively (negatively) invariant* if $T(H) \subset H$ ($H \subset T(H)$). $H$ is said to be *invariant* if $T(H) = H$.

We shall show in a moment that the limit set of a bounded motion is closed and invariant. We cannot expect, as is the case for continuous motions, that it be connected. However, relative to invariance it does have a connectedness property.

**4.2.** DEFINITION. A closed invariant set $H$ is said to be *invariantly connected* if it is not the union of two nonempty disjoint closed invariant sets.

**4.3.** DEFINITION. A motion $T^n x$ is said to be *periodic* (or *cyclic*) if for some $k > 0$, $T^k x = x$. The least such integer $k$ is called the *period* of the motion or the *order* of the cycle. If $k = 1$, $x$ is a fixed point of $T$ and is called an *equilibrium state of* (1.1).

**4.4.** *Exercise.* Show that: An invariant set with a finite number of elements is invariantly connected if and only if it is a periodic motion.

**4.5.** *Exercise.* Show that:

(a) The closure of a positively invariant set is positively invariant.

(b) The closure of a bounded invariant set is invariant.

**4.6.** *Exercise.* Give an example of an invariant set whose closure is not invariant.

**4.7.** DEFINITION. $T_n x$ (defined for all $n \in J$) is called an *extension of the motion* $T^n x$ if $T_0 x = x$ and $T(T_n x) = T_{n+1} x$ for all $n \in J$.

N.B. $T_n x = T^n x$ for all $n \in J_+$ and, if $x$ is contained in an invariant set $H$, the motion $T^n x$ always has an extension, which may or may not be unique.

**4.8.** *Exercise.* Show that a set $H$ is invariant if and only if each motion starting in $H$ has an extension in $H$ for all $n$.

**4.9.** *Exercise.* Give an example to show that an invariant set $H$ can have a motion extended from a point in $H$ that is not in $H$.

**4.10.** *Exercise.* Let $E$ be a given set in $R^m$ and let $M$ be the largest (by inclusion) invariant set in $E$. Show that:

(a) $M$ is the union of all extended motions that remain in $E$ for all $n \in J$.
(b) $x \in M$ if and only if there is an extended motion $T_n$ with $T_n x \in E$ for all $n \in J$.
(c) If $E$ is compact, then $M$ is compact.

**5. Basic properties of limit sets.** In the next section we shall both extend and unify Liapunov's direct method by exploiting the basic properties of limit sets.

**5.1.** THEOREM. *Every limit set $\Omega(x)$ is closed and positively invariant.*

*Proof.* It is easy to see that the complement of $\Omega(x)$ is open, and hence $\Omega(x)$ is closed. Suppose $y \in \Omega(x)$. Then there is a sequence of integers $n_i$ such that $n_i \to \infty$ and $T^{n_i} x \to y$ as $i \to \infty$. By the continuity of $T$, $T(T^{n_i} x) = T^{n_i+1} x \to Ty$, and $Ty \in \Omega(x)$. Hence $T(\Omega(x)) \subset \Omega(x)$, and $\Omega(x)$ is positively invariant. ∎

What we are most interested in is the asymptotic behavior of bounded motions. A motion $T^n x$ which is bounded for all $n \in J_+$ is often said to be *positively stable in the sense of Lagrange.*

**5.2.** THEOREM. *If $T^n x$ is bounded for $n \in J_+$, then $\Omega(x)$ is nonempty, compact, invariant, invariantly connected, and is the smallest closed set that $T^n x$ approaches as $n \to \infty$.*

*Proof.* The boundedness of $T^n x$ clearly implies that $\Omega(x)$ is nonempty and bounded, and hence by Theorem 5.1 is compact. Let $y$ be in $\Omega(x)$ and select $n_i$ as in the proof of Theorem 5.1. By boundedness of $T^n(x)$, we may assume that $T^{n_i-1} x$ also converges (by selecting a subsequence, if necessary). Let $T^{n_i-1} x \to z$. Then $z \in \Omega(x)$, $T(T^{n_i-1} x) = T^{n_i} x \to Tz = y$. Therefore $\Omega(x) \subset T(\Omega(x))$ and, by Theorem 5.1, $\Omega(x)$ is invariant.

We shall show next that $T^n x \to \Omega(x)$ when $T^n x$ is bounded. Since $\rho(T^n(x), \Omega(x))$ is bounded, we conclude that, if $T^n(x)$ does not approach $\Omega(x)$, there is a sequence $n_i$ such that $n_i \to \infty$, $T^{n_i}(x)$ converges, and $\rho(T^{n_i} x, \Omega(x))$ does not approach 0 as

$i \to \infty$. Clearly this is a contradiction, since the limit of $T^{n_i}x$ is in $\Omega(x)$, and $T^n x \to \Omega(x)$ as $n \to \infty$. If $T^n x \to E$ as $n \to \infty$ and $E$ is closed, then clearly $\Omega(x) \subset E$. Hence $\Omega(x)$ is the smallest closed set that $T^n x$ approaches as $n \to \infty$.

It remains to show that $\Omega(x)$ is invariantly connected. Assume that $\Omega(x)$ is the union of two disjoint closed nonempty invariant sets $\Omega_1$ and $\Omega_2$. Since $\Omega(x)$ is compact, so are $\Omega_1$ and $\Omega_2$. There then exist disjoint open sets $U_1$ and $U_2$ such that $\Omega_1 \subset U_1$ and $\Omega_2 \subset U_2$. Also, since $T$ is continuous and therefore uniformly continuous on $\Omega_1$, there is an open set $V_1$ such that $\Omega_1 \subset V_1$ and $T(V_1) \subset U_1$. Since $\Omega(x)$ is the smallest closed set that $T^n(x)$ approaches, $T^n(x)$ must intersect both $V_1$ and $U_2$ an infinite number of times. But this implies the existence of a convergent subsequence $T^{n_i}x$ that is not in either $V_1$ or $U_2$. Since $\Omega(x)$ is in the union of $V_1$ and $U_2$, this is a contradiction, and $\Omega(x)$ is invariantly connected. ■

**5.3.** *Exercise.* Give an example of a limit set $\Omega(x)$ with the property that $\Omega(x)$ is nonempty and $T^n x$ does not approach $\Omega(x)$ as $n \to \infty$.

**5.4.** *Exercise.* What changes can be made in the above results if in Definition 2.1 $J_+$ is replaced by $J$?

**5.5.** *Exercise* (see Exercise 3.3). Establish the analogous basic properties of $\Omega(H)$.

**5.6.** *Exercise.* Show that: If $K$ is compact and positively invariant, then $\Omega(K) = \bigcap_{n=0}^{\infty} T^n(K)$ is nonempty, compact, and invariant, and is obviously the largest invariant set in $K$. (This is a special case of Exercise 5.5).

**6. Liapunov functions. An extension of Liapunov's direct method.** Information about the location of the limit set of a motion is information about its asymptotic behavior. For instance, if we knew that a motion had to approach a set with a finite number of elements, we would know by Theorem 5.2 and Exercise 4.4 that the motion approaches a periodic motion. What we shall do here is to show that, suitably defined, Liapunov functions give information about the location of limit sets. This is done exploiting, in particular, the invariance property of limit sets, and for this reason the idea behind what we are about to do is called the "invariance principle." It is, as we shall see, an elementary and simple idea, but it has proved to be useful both for theory and applications and to be capable of considerable generalization (see Chapter 3 and Appendix A).

Let $V: R^m \to R$. Relative to (1.1) (or to $T$) define

$$\dot{V}(x) = V(T(x)) - V(x). \tag{6.1}$$

If $x(n)$ is a solution of (1.1),

$$\dot{V}(x(n)) = V(x(n+1)) - V(x(n)),$$

and $\dot{V}(x) \leq 0$ means that $V$ is nonincreasing along solutions. Computing $\dot{V}(x)$ does not require a knowledge of solutions—it is computed directly from a knowledge of the right-hand side of (1.1)—which is why the resulting method is called "direct."

**6.1.** DEFINITION. Let $G$ be any set in $R^m$. We say that $V$ is a *Liapunov function of* (1.1) *on* $G$ if (i) $V$ is continuous and (ii) $\dot{V}(x) \leqq 0$ for all $x \in G$.

Note that in this definition we could replace (ii) by the condition that $\dot{V}$ not change sign in $G$.

**6.2. Notation.** For $V$ a Liapunov function of (1.1) on $G$, we define

$$E = \{x;\ \dot{V}(x) = 0,\ x \in \bar{G}\}. \tag{6.2}$$

We use $M$ to denote the largest invariant set in $E$, and $V^{-1}(c) = \{x;\ V(x) = c,\ x \in R^m\}$.

**6.3.** THEOREM (invariance principle). *If* (i) $V$ *is a Liapunov function of* (1.1) *on* $G$, *and* (ii) $x(n)$ *is a solution of* (1.1) *bounded and in* $G$ *for all* $n \geqq 0$, *then there is a number* $c$ *such that* $x(n) \to M \cap V^{-1}(c)$ *as* $n \to \infty$.

*Proof.* Let $x^0 = x(0)$ so that $x(n) = T^n x_0$. Now our assumptions imply that $V(x(n))$ is nonincreasing with $n$ and is bounded from below, and hence $V(x(n)) \to c$ as $n \to \infty$. Let $y \in \Omega(x^0)$. Then there is a sequence $n_i$ such that $n_i \to \infty$ and $x(n_i) \to y$. Since $V$ is continuous, $V(x(n_i)) \to V(y) = c$, and $\Omega(x^0) \subset V^{-1}(c)$. Since $\Omega(x^0)$ is invariant, $V(Ty) = c$ and $\dot{V}(y) = 0$. Therefore $\Omega(x^0) \subset E$, and hence is in $M$. Since $x(n) \to \Omega(x^0)$, $x(n) \to M \cap V^{-1}(c)$. ■

We now look at a simple example to illustrate how the result can be applied. Later we shall consider some special corollaries of this basic result.

**6.4.** *Example.* Consider the 2-dimensional system

$$x(n+1) = \frac{ay(n)}{1+x^2(n)},$$

$$y(n+1) = \frac{bx(n)}{1+y^2(n)}$$

or

$$x' = \frac{ay}{1+x^2},$$

$$y' = \frac{bx}{1+y^2}. \tag{6.3}$$

Take $V(x, y) = x^2 + y^2$. Then

$$\dot{V}(x, y) = \left(\frac{b^2}{(1+y^2)^2} - 1\right)x^2 + \left(\frac{a^2}{(1+x^2)^2} - 1\right)y^2.$$

*Case* 1. $a^2<1$, $b^2<1$. Now

$$\dot{V} \leqq (b^2-1)x^2+(a^2-1)y^2,$$

and $V$ is a Liapunov function of (6.3) on $R^2$. Here $M=E=\{(0, 0)\}$, and since every solution is clearly bounded, we have by Theorem 6.3 that every solution approaches the origin as $n \to \infty$ (the origin is a global attractor and, as we shall explain later, we can conclude in this case that the origin is globally asymptotically stable). This is Liapunov's classical case—$V(x)$ and $-\dot{V}(x)$ are positive definite.

*Case* 2. $a^2 \leqq 1, b^2 \leqq 1$ and $a^2+b^2<2$. We may assume that $a^2<1$ and $b^2=1$. $V$ is still a Liapunov function of (6.3) on $R^2$ but $-\dot{V}$ is not positive definite. In fact, $\dot{V} \leqq (a^2-1)y^2$, and $E$ is the $x$-axis ($y=0$). However, since $T(x, 0)=(0, bx)$, we see that again $M$ is just the origin, and we can conclude that the origin is globally asymptotically stable.

*Case* 3. $a^2=b^2=1$. $V$ is still a Liapunov function of (6.3) and all solutions are still bounded. Here $E=M$ is the union of the two coordinate axes, and by Theorem 6.3 we know that each solution approaches $\{(c, 0), (0, c), (-c, 0), (0, -c)\}$ for some $c$—the intersection of $E$ with the circle $x^2+y^2=c^2$. There are two subcases.

(i) $ab=1$. Then $T(c, 0)=(0, bc)$, $T^2(c, 0)=T(abc, 0)=(c, 0)$. Since limit sets are invariantly connected, every solution approaches one of these periodic motions—the origin or a periodic motion of period 2 (see Exercise 4.4).

(ii) $ab=1$. Here $T(c, 0)=(0, bc)$, $T^2(c, 0)=(abc, 0)=(-c, 0)$, $T^3(c, 0)=(0, -bc)$, and $T^4(c, 0)=(-abc, 0)=(c, 0)$. If $c \neq 0$, these are periodic motions of period 4. As in (i) above, each solution approaches the origin or one of these periodic motions of period 4.

*Case* 4. $a^2>1$, $b^2>1$. Let $B_\delta=\{(x, y); x^2+y^2<\delta^2\}$. For $x \in B_\delta$ and $\delta$ sufficiently small,

$$\dot{V} \geqq \left(\frac{b^2}{1+\delta^2}-1\right)x^2+\left(\frac{a^2}{1+\delta^2}-1\right)y^2 \geqq 0,$$

and $-V$ is a Liapunov function of (4.2) on $B_\delta$ for $\delta$ sufficiently small, and $E=M=\{(0, 0)\}$. No solution starting at a point in $B_\delta$ other than the origin can approach the origin from within $B_\delta$ (its distance from the origin is increasing) and $T(x, y)=(0, 0)$ implies $x=y=0$. Therefore each such solution must leave $B_\delta$ by Theorem 6.3 (instability) and, since no solution can jump to the origin in finite time except the trivial solution, there is no nontrivial solution that can approach the origin as $n \to \infty$.

**7. Stability and instability.** We shall define here the concepts of stability and instability for sets relative to the basic difference equation (1.1) or, in other words, relative to $T$.

**7.1.** DEFINITION. A set $H$ is said to be *stable*, if given a neighborhood $U$ of $H$ (an open set containing $\bar{H}$), there is a neighborhood $W$ of $H$ such that $T^n(W) \subset U$ for all $n \in J_+$.

The next exercise shows that we might as well have restricted ourselves to closed positively invariant sets.

**7.2.** *Exercise.* Show that: If $H$ is stable, then $\bar{H}$ is positively invariant. In particular, if a point is stable, it is an equilibrium point.

**7.3. Notation.** For $H$ a set in $R^m$ define $\hat{H}$ as follows: $x \in \hat{H}$ if there exist sequences $x_i \in R^m$ and $n_i \in J_+$ such that $x_i \to y \in \bar{H}$ and $T^{n_i}(x_i) \to z$.

In topological dynamics $\hat{H}$ is called the *prolongation* of $H$. The set of all such $z$ for which $n_i \to \infty$ as $i \to \infty$ is called the *prolongation limit set* of $H$. Note that $H \subset \hat{H}$. The properties of $\hat{H}$ that concern us here are given in the following lemma.

**7.4.** LEMMA.

(i) *Let $H$ be a compact positively invariant set. Then $H$ is stable if and only if $\hat{H} = H$.*

(ii) *Let $H$ be a closed invariant set contained in an open bounded positively invariant set $G$. Then $\hat{H}$ is invariant.*

*Proof.*

(i) Clearly, if there is a $z \in \hat{H}$ that is not in $H$, then $H$ is not stable. Conversely, suppose $H$ is not stable. Then for some neighborhood $U$ of $H$, which we may assume is bounded, there is a sequence $x^i \in U$ such that $x^i \to y \in H$ as $i \to \infty$ and each motion $T^n x^i$ eventually leaves $U$. Let $n_i$ be the smallest integer with the property that $T^{n_i} x^i$ is not in $U$. Now $T^{n_i} x^i \in T(U)$, and this sequence is bounded. Since it contains a convergent subsequence whose limit is not in $H$, $H$ not stable implies $\hat{H} \neq H$. ■

(ii) Let $z \in \hat{H}$. Then there exist sequences $x_i \in G$ and $n_i \in J_+$ such that $x_i \to y \in H$ and $T^{n_i} x_i \to z$ as $i \to \infty$. Since $T^{n_i+1} x_i \to Tz$, we see that $Tz \in \hat{H}$ and $T(\hat{H}) \subset \hat{H}$. If $n_i$ is bounded, there is a $k \in J_+$ with $T^k x_i \to z$. But then $z = T^k y \in H$, since $H$ is invariant, and consequently there is a $w \in H \subset \hat{H}$ with $Tw = z$. If $n_i$ is not bounded, we may assume that $n_i \to \infty$ as $i \to \infty$. Now $T^{n_i - 1} x_i$ is in $G$, is therefore bounded, and we may assume that $T^{n_i - 1} x_i \to w$. Then $w \in \hat{H}$, and again we have the existence of $w \in \hat{H}$ with $Tw = z$. This proves that $\hat{H} \subset T(\hat{H})$, and $\hat{H}$ is invariant. ■

The type of stability of greatest importance in applications is "asymptotic stability". We shall see why in § 11.

**7.5.** DEFINITION. A set $H$ is an *attractor* if there is a neighborhood $U$ of $\bar{H}$ such that $x \in U$ implies $T^n x \to \bar{H}$ as $n \to \infty$. $H$ is said to be *asymptotically stable* if it is both stable and an attractor. If, in addition, $T^n x \to \bar{H}$ as $n \to \infty$ for all $x \in R^m$, $H$ is said to be *globally asymptotically stable*. *Unstable* means not stable. If $H$ is neither stable nor an attractor it will be said to be *strongly unstable*.

**7.6.** *Exercise.* Given a set $H$ its inverse image is $T^{-1}(H) = \{z\,;\, T(z) \in H\}$. A set is said to be *inversely invariant* if $T^{-1}(H) = H$.

Show that:

(a) A set $H$ is inversely invariant if and only if $H$ is positively invariant and $T^{-1}(H) \subset H$.

(b) A set $H$ is negatively invariant if and only if $T^{-1}(x)$ intersects $H$ for each $x \in H$.

**7.7.** *Exercise.* The *region of attraction* $\mathcal{R}(H)$ of a set $H$ is the set of all $x$ such that $T^n x \to H$ as $n \to \infty$. The boundary of $H$ is denoted by $\partial H$ and its complement by $\mathcal{C}H$. Show that: If $H$ is asymptotically stable, then

(a) $\mathcal{R}(H)$ is open.

(b) $\mathcal{R}(H)$, $\partial\mathcal{R}(H)$ and $\mathcal{C}\mathcal{R}(H)$ are inversely invariant.

**7.8.** *Exercise.* Give an example of an asymptotically stable set $H$ for which $\mathcal{R}(H)$, $\partial\mathcal{R}(H)$ and $\mathcal{C}\mathcal{R}(H)$ are each not invariant.

**7.9.** THEOREM. *Let $G$ be a bounded open positively invariant set. If* (i) *$V$ is a Liapunov function of* (1.1) *on $G$, and* (ii) *$M \subset G$, then $M$ is an attractor and $\bar{G} \subset \mathcal{R}(M)$. If, in addition,* (iii) *$V$ is constant on $M$, then $M$ is asymptotically stable (globally asymptotically stable relative to $G$).*

*Proof.* Since $V$ and $T$ are continuous, $\dot{V}$ is continuous and $E$ is closed. Now $M$ is the largest invariant set in $E$, and therefore is closed (see Exercise 4.5(b)). Hence by Theorem 6.3 we conclude that $M$ is an attractor. Also the continuity of $\dot{V}$ implies that $V$ is a Liapunov function on $\bar{G}$. By Exercise 4.5(a) $\bar{G}$ is positively invariant, and by Theorem 6.3 we see that $\bar{G} \subset \mathcal{R}(M)$. Note that $\hat{M} \subset \bar{G}$. It remains to show that $M$ is stable when $V(x) = c$ on $M$. We do this by showing that $\hat{M} \subset E$. Then, since $\hat{M}$ in invariant (Lemma 7.4) and $M$ is the largest invariant set in $E$, it follows that $\hat{M} = M$, and $M$ is stable. Let $z \in \hat{M}$. Then $T^n z \to M$ and $V(z) \geqq c$. Now there exist $x_i$ and $n_i$ such that $x_i \to y \in M$ and $T^{n_i} x_i \to z$ as $i \to \infty$. Since $V(x_i) \geqq V(T^{n_i} x_i)$, we see that $c \geqq V(z)$, and $V(z) = c$ for each $z \in \hat{M}$. Now $\hat{M}$ is invariant, and therefore $\hat{M} \subset E$. ■

Note that condition (iii) of Theorem 7.9 is automatically satisfied if $M$ is a single point or if $M$ is an invariantly connected set with a finite number of elements. Thus we have obtained a sufficient condition for asymptotic stability without assuming that either $V$ or $-\dot{V}$ is positive definite with respect to $M$. This is how we were able to conclude asymptotic stability in Example 6.4. Also this result shows clearly that information is obtained on the "extent" of asymptotic stability since one then knows that the region of asymptotic stability is larger than $\bar{G}$. Now it usually turns out in applications that $M$ is also the largest positively invariant set in $E$, and hence that $V(x) - c$ is positive definite relative to $M$. Thus, usually, a "good" Liapunov function will be positive definite but this result says it need not be verified. A failure to recognize this can cause, and often has caused, a lot of unnecessary work. Even when $M$ is a single point positive definiteness can be difficult to establish. Actually, one can look upon Theorem 7.9 and Exercise 7.10 as a sufficient condition for positive definiteness.

**7.10.** *Exercise.* Show that: If (i), (ii), and (iii) of Theorem 7.9 are satisfied and $M$ is also the largest positive invariant set in $E$, then $V(x)>c$ for $x \in G-M$, where $c$ is the value of $V$ on $M$; i.e., $V(x)-c$ is positive definite relative to $M$.

**7.11.** *Exercise.* (*The analogue of Liapunov's theorems on stability and asymptotic stability.*) Let $H$ be compact and let $G$ be an open set containing $H$.

Show that:

(i) $V(x) \leqq 0$ for $x \in H$ and $V(x)>0$ for $x \in G-H$, and

(ii) $V$ is a Liapunov function of (1.1) on $G$,

then $H$ is stable.

If, in addition,

(iii) $E \subset H$,

then $H$ is asymptotically stable.

**7.12.** *Remark.* Condition (i) says that $V$ is positive definite relative to $H$, and conditions (ii) and (iii) imply that $\dot{V}$ is negative definite relative to $H$. We know that condition (iii) can be replaced by $M \subset H$. Also, if $V$ is constant on $M$, it is not necessary to assume $V$ is positive definite.

**7.13.** *Example.* Using the above results, we can now conclude, when $a^2 \leqq 1$ and $b^2 \leqq 1$ and $a^2+b^2<2$ in Example 6.4, that the origin is globally asymptotically stable. Note that $\dot{V}$ is not negative definite.

**7.14.** *Exercise.* Show that: If no solution (motion) outside $M$ in $G$ can reach $M$ in finite time ($M$ is also inversely invariant relative to $G$), then condition (iii) in Theorem 7.9 can be replaced by: (iii)′ *$V$ is constant on the boundary of $M$.*

We now give for difference equations the analogue and an extension of Ĉetaev's theorem on instability for ordinary differential equations. Ĉetaev wanted to show for conservative dynamical systems that, if an equilibrium is not a minimum of the potential energy, then it is not stable (see [62, p. 56]).

**7.15.** THEOREM (an instability theorem). *Let $y$ be an equilibrium point on the boundary of an open set $U$, and let $N$ be a neighborhood of $y$. Assume that*: (i) *$V$ is a Liapunov function of* (1.1) *on* $G=U \cap N$, (ii) *$M \cap G$ is empty,* (iii) $V(x)=c$ *on that part of the boundary of $U$ inside $N$, and* (iv) $V(x)<c$ *for $x \in G$. If $T(G) \subset U$, then $y$ is unstable; in fact, if $N_0$ is any bounded neighborhood of $y$ properly contained in $N$, then each solution starting inside $G_0=U \cap N_0$ eventually leaves $N_0$. If $U$ is positively invariant, then $y$ is strongly unstable; in fact, no solution starting in $U$ can approach $y$ as $n \to \infty$.*

*Proof.* Any solution starting in $G_0$ must by Theorem 6.3 eventually leave $G_0$ since it cannot approach that part of the boundary of $U$ inside $N$. Since its first exit from $G_0$ is still in $U$, it must leave $N_0$. Since $y$ is on the boundary of $G_0$, $y$ is not stable. If $U$ is positively invariant, then it is clear that $y$ cannot be an attractor. ■

The following corollary of the above theorem contains the analogues of Liapunov's first two instability theorems.

**7.16.** COROLLARY. *Let $V: R^m \to R$ be continuous and $T(0)=0$. Arbitrarily near the origin there are points where $V(x)>0$, and $V(0)=0$. In a neighborhood $N$ of the origin and relative to* (1.1)

$$\dot{V}(x)=\beta V(x)+W(x),$$

*where either* (i) *$W(x)$ is non-negative and $\beta>0$ or* (ii) *$\beta \geqq 0$ and $W$ is positive definite. Then the origin is unstable.*

*Proof.* Let $U$ correspond to $V(x)>0$. Then $-V$ is a Liapunov function of (1.1) on $G=N \cap U$. Since $x \in G$ implies $T(x) \in U$, the instability of the origin follows from Theorem 7.15. ■

**7.17.** *Example.* Consider

$$x_1' = g_1(x_1, x_2),$$
$$x_2' = g_2(x_1, x_2), \quad g_1(0,0)=g_2(0,0)=0.$$

Let $V=-x_1x_2$. Then $\dot{V}=x_1x_2-g_1(x)g_2(x)$. Suppose:

(a) $g_1(x)g_2(x)>0$ for $x_1x_2>0$, and

(b) $g_1(x)g_2(x)>x_1x_2$ for $x_1x_2>0$ and $x$ sufficiently close to the origin (within a neighborhood $N$ of the origin).

Then with $G$ corresponding to $x_1x_2>0$ the conditions of Theorem 7.15 are satisfied, and the origin is strongly unstable.

**7.18.** *Exercise.* State and prove a theorem similar to Theorem 7.15 when $y$ is in $U$.

**8. Vector Liapunov functions.** Vector Liapunov functions for ordinary differential equations were first discussed and illustrated in [12]. (For an application and further references, see [30].) Here we present for difference equations the results of [60]. The usual generalization from scalar to vector Liapunov functions is straightforward. Scalar inequalities are replaced by vector inequalities, and there is nothing essentially new. It is a notionally convenient means of taking advantage of the fact that the sum of Liapunov functions is a Liapunov. We shall present, based on an idea introduced by the economists Arrow, Block and Hurwicz in [4], another, less obvious, way of constructing a scalar Liapunov function from a number of scalar functions.

**8.1. Notation.** For $x \in R^m$:

$x>0$ means $x_i>0$ for $i=1, \cdots, m$,

$x \geqq 0$ means $x_i \geqq 0$ for $i=1, \cdots, m$,

$\underline{R_+^m}=\{x; x>0, x \in R^m\}$,

$R_+^m=\{x; x \geqq 0, x \in R^m\}$.

Similarly for matrices, $B=(b_{ij})>0$ $(B \geqq 0)$ means $b_{ij}>0$ $(b_{ij} \geqq 0)$, and $|B|=(|b_{ij}|)$.

Let $v: R^m \to R^q$ and define

$$\dot{v}(x)=v(Tx)-v(x).$$

**8.2.** DEFINITION. Let $G$ be any set in $R^m$. We say that $v$ is a *vector Liapunov function of* (1.1) *on* $G$ if (i) $v$ is continuous on $R^m$, and (ii) $\dot{v}(x) \leqq 0$ for all $x \in G$. Define $E = \{x; \dot{v}(x) = 0, x \in \bar{G}\}$ and let $M$ be the largest invariant set in $E$.

Then, just as for scalar Liapunov functions, we have the following theorem.

**8.3.** THEOREM. *Let $v$ be a vector Liapunov function of* (1.1) *on $G$. If $x(n)$ is a solution of* (1.1) *that is bounded and in $G$ for all $n \in J_+$, then there is a $c \in R^q$ such that $x(n) \to M \cap v^{-1}(c)$ as $n \to \infty$.*

**8.4.** *Example.* Consider the linear difference equation (*) $x' = Ax$, where $A \geqq 0$, $A$ an $m \times m$ matrix. Assume also that $(I-A)^{-1}$ exists and $(I-A)^{-1} \geqq 0$. Let $v(x) = (I-A)^{-1}x$. Then $\dot{v}(x) = -x$, and we see that $v(x)$ is a Liapunov function of (*) on $R^m_+$. Now $R^m_+$ is positively invariant, and each solution starting in $R^m_+$ is bounded since, for each $c \in R^m$, $G_c = \{x; v(x) \leqq c, x \geqq 0\}$ is bounded. Now $M$ is the origin and therefore by Theorem 8.3 every solution of (*) starting in $R^m_+$ approaches the origin as $n \to \infty$. Since $R^m_+$ contains a basis of $R^m$, every solution of (*) approaches 0 as $n \to \infty$, and the origin is globally asymptotically stable.

To say that $v$ is a vector Liapunov function of (1.1) on $G$ means, of course, that each component $v_i$ of $v$ is a scalar Liapunov function of $G$ with an associated $E_i$ and $M_i$. Then $V = \sum_{i=1}^q v_i$ is a scalar Liapunov function of (1.1) on $G$, and the sets $E$ and $M$ associated with $V$ are the same as the sets $E$ and $M$ in Definition 8.2 associated with the vector Liapunov function $v$. However, as Example 8.4 illustrates, there can be a notational advantage to the use of a vector Liapunov function.

Generalizing an idea by Arrow, Block and Hurwicz in [4] for the construction of a Liapunov function, we shall now introduce a concept of a vector Liapunov function that does not require any of its components to be a scalar Liapunov function.

**8.5.** DEFINITION. Let $w: R^m \to R^q$ and define $W(x) = \max_i w_i(x)$ and $\mathring{w}(x) = w(T(x)) - W(x)u$, where $u_i = 1, i = 1, 2, \cdots, q$, $(\mathring{w}_i(x) = w_i(Tx) - W(x))$. We shall say that $w$ is a *vector Liapunov function of* (1.1) *on* $G$ if (i) $w$ is continuous on $R^m$, and (ii) $\mathring{w}(x) \leqq 0$ for all $x \in G$. Define $E = \{x; w(x) \nless 0, x \in \bar{G}\}$, and $M$ is the largest invariant set in $E$.

Note that $x \in E$ if $x \in \bar{G}$ and, for some $i = 1, \cdots, q$, $\mathring{w}_i(x) = 0$. Also

$$\mathring{w}(x) = \dot{w}(x) + w(x) - W(x)u$$

so that $\mathring{w}(x) \leqq \dot{w}(x)$ and requiring $\mathring{w}(x) \leqq 0$ is not as stringent as requiring $\dot{w}(x) \leqq 0$. In fact, $w$ can be a vector Liapunov function on $G$ in this sense, and yet it may be that no component of $w(x)$ is a scalar Liapunov function on $G$. However, $\dot{W}(x) = \max_i \mathring{w}_i(x)$ so that, if $w$ is a vector Liapunov function on $G$, $W$ is a scalar Liapunov function on $G$. We then have immediately the following.

**8.4.** THEOREM. *If* (i) *$w$ is a vector Liapunov function of* (1.1) *on $G$ and* (ii) *a solution $x(n)$ of* (1.1) *is in $G$ and bounded for all $n \geqq 0$, then, for some number $c$, $x(n) \to M \cap W^{-1}(c)$ as $n \to \infty$.*

**8.5.** *Example.* Consider again the linear difference equation $x' = Ax$. Assume there is a $c > 0$ in $R^m$ such that $|A|c < c$. Let $w_i(x) = |x_i|/c_i$ so that $W(x) = \max_i |x_i|/c_i$. Then

$$w_i(Ax) = \frac{|(Ax)i|}{c_i} \leqq \frac{1}{c_i} \sum_{k=1}^{m} |a_{ik}| c_k \frac{|x_k|}{c_k}$$

$$\leqq W(x) \frac{(|A|c)_i}{c_i} < W(x), \qquad x \neq 0.$$

Hence $\mathring{w}(x) < 0$ for $x \neq 0$ and $w(x)$ is a vector Liapunov function on $R^m$. $M$ is the origin. Since $W(x) \to \infty$ as $|x| \to \infty$, every solution is bounded, and hence approaches the origin as $n \to \infty$. Therefore the existence of a $c > 0$ such that $|A|c < c$ is a sufficient condition for global asymptotic stability of the origin.

**8.6.** *Exercise.* Show that: The intersection of positively invariant sets is positively invariant.

**8.7.** *Exercise.* Show that: If $T$ is one-to-one, the intersection of invariant sets is invariant.

**8.8.** *Exercise.* Give an example of the intersection of invariant sets that is not invariant.

**8.9.** *Exercise.* Assume that $v_i$, $i = 1, \cdots, q$, are scalar Liapunov functions of (1.1) on $G$ with $E_i = \{x; \dot{v}_i(x) = 0, x \in \bar{G}\}$ and $M_i$ the largest invariant set in $E_i$. Define $E = \bigcap_{i=1}^{q} E_i$ and $M^* = \bigcap_{i=1}^{q} M_i$. Let $M$ be the largest invariant set in $E$ and let $M^0$ be the largest invariant set in $M^*$. Show that $M = M^0$.

**8.10.** *Remark.* Assume that $v$ is a vector Liapunov function on $G$ in the sense of Definition 8.2. Then each $v_i$ is a scalar Liapunov function on $G$. The point of Exercise 8.9 is that if $x(n)$ is a solution of (1.1) bounded and in $G$ for all $n \in J_+$, then by Theorem 6.3 we know that, for some number $c_i$, $x(n) \to M_i \cap v_i^{-1}(c_i)$ as $n \to \infty$ for each $i = 1, \cdots, q$; i.e., $x(n) \to M^* \cap v^{-1}(c)$, where $c = (c_1, c_2, \cdots, c_q)$ and $v = (v_1, v_2, \cdots, v_q)$. Hence $x(n) \to M^0 \cap v^{-1}(c)$ as $n \to \infty$. But by Exercise 8.9 this observation yields nothing new.

## 9. Linear difference equations

**9.1. Notation.**

$A = (a_{ij})$ is a real $m \times m$ matrix,
$\sigma(A) = \{\lambda_1, \lambda_2, \cdots, \lambda_m\}$ is the set of eigenvalues of $A$ (the *spectrum* of $A$),
$r(A) = \max_i |\lambda_i|$ (the *spectral radius*),
$Rl\,\sigma(A) = \max_i Rl\,\lambda_i$,
$A^T = (a_{ji})$, the *transpose* of $A$,
$A^n = AA^{n-1}$, $A^0 = I$.

The general linear autonomous difference equation of dimension $m$ is

$$x' = Ax. \tag{9.1}$$

The solution satisfying $x(0) = x^0$ is $A^n x^0$. The columns of $A^n$ are the *principal solutions* of (9.1). If $v^i$ is an eigenvector of $A$ with eigenvalue $\lambda_i$, then $c_i \lambda_i^n v^i$ is a solution of (9.1). Thus, if $r(A) \geqq 1$, there is aways a solution that does not approach the origin. If $r(A) > 1$, there are unbounded solutions. If the eigenvalues of $A$ are distinct, then the general solution of (9.1) is

$$x(n) = c_1 \lambda_1^n v^1 + \cdots + c_n \lambda_m^n.$$

**9.2. An algorithm for computing $A^n$ from its eigenvalues.** This algorithm is the analogue of Putzer's algorithm in [93] for computing $e^{At}$.

We look for a representation of $A^n$ in the form

$$A^n = \sum_{j=1}^{m} w_j(n) Q_{j-1}, \tag{9.2}$$

where

$$Q_j = (A - \lambda_j I) Q_{j-1}, \qquad Q_0 = I. \tag{9.3}$$

$Q_m = 0$ by the Hamilton–Cayley theorem (every matrix satisfies its characteristic equation). It is just this fact that suggests the form of the representation (9.2).

The initial condition $A^0 = I$ is satisfied by taking $w_1(0) = 1$, $w_2(0) = \cdots = w_m(0) = 0$. We want $A \sum_{j=1}^m w_n(n) Q_{j-1} = \sum_{j=1}^m w_j(n+1) Q_{j-1}$ or, since $AQ_{j-1} = Q_j + \lambda_j Q_{j-1}$,

$$\sum_{j=1}^{m} w_j(n)(Q_j + \lambda Q_{j-1}) = \sum_{j=1}^{m} w_j(n+1) Q_{j-1}.$$

Thus, (9.2) holds if

$$\begin{aligned} &w_1(n+1) = \lambda_1 w_1(n), \quad w_1(0) = 1 \quad (w_1(n) = \lambda_1^n), \\ &w_j(n+1) = \lambda_j w_j(n) + w_{j-1}(n), \quad w_j(0) = 0, \qquad j = 2, \cdots, m. \end{aligned} \tag{9.4}$$

Equations (9.3) and (9.4) are algorithms for computing the $Q_j$ and the $w_j(n)$ in terms of the eigenvalues of $A$ (i.e., for computing $A^n$ if we know or have computed the eigenvalues of $A$).

By way of illustration let us use the algorithm to find the solution of

$$y''' - 3y'' + 3y' - y = 0, \quad y''(0) = y_0'', \quad y'(0) = y_0', \quad y(0) = y_0.$$

This third-order equation is equivalent to $x' = Ax$, where

$$x = \begin{pmatrix} x_1 \\ x_2 \\ x_3 \end{pmatrix} = \begin{pmatrix} y \\ y' \\ y'' \end{pmatrix} \quad \text{and} \quad A = \begin{pmatrix} 0 & 1 & 0 \\ 0 & 0 & 1 \\ 1 & -3 & 3 \end{pmatrix},$$

$$\phi(\lambda) = \det(A - \lambda I) = -(\lambda - 1)^3 \quad \text{and} \quad \lambda_1 = \lambda_2 = \lambda_3 = 1,$$

$$Q_0=I,\quad Q_1=A-I=\begin{pmatrix}-1 & 1 & 0\\ 0 & -1 & 0\\ 1 & -3 & 2\end{pmatrix},$$

$$Q_2=(A-I)^2=\begin{pmatrix}1 & -2 & 1\\ 1 & -2 & 1\\ 1 & -2 & 1\end{pmatrix}.$$

Solving (9.4) directly, or by using Exercise 9.4, we obtain $w_1(n)=1$, $w_2(n)=n$, $w_3(n)=\frac{1}{2}n(n-1)$. Hence,

$$\begin{aligned}A^n&=I+n(A-I)+\tfrac{1}{2}n(n-1)(A-I)^2\\ &=\begin{pmatrix}\frac{1}{2}(n-1)(n-2) & -n(n-2) & \frac{1}{2}n(n-1)\\ \frac{1}{2}n(n-1) & -(n+1)(n-1) & \frac{1}{2}(n+1)n\\ \frac{1}{2}(n+1)n & -(n+2)n & \frac{1}{2}(n+2)(n+1)\end{pmatrix}.\end{aligned}$$

The solution $y(n)$ is the first component of $A^nx^0$. This gives $y(n)=\frac{1}{2}(n-1)(n-2)y_0-n(n-2)y_0'+\frac{1}{2}n(n-1)y_0''$.

**9.3.** *Exercise.* (a) Show that: If $r(A)=r_0<\beta$, then

$$|w_j(n)|\leqq\frac{\beta^n}{(\beta-r_0)^{j-1}},\qquad j=1,\cdots,m.$$

(b) If $\alpha<\min_i|\lambda_i|$, establish a lower bound for $|w_j(n)|$.

From Exercise 9.3 we see that, if $|r(A)|<1$, then $w_j(n)\to 0$ as $n\to\infty$, and hence $A^n\to 0$ as $n\to\infty$. We already know that when $r(A)\geqq 1$, $A^n$ does not approach 0 as $n\to\infty$. $A^n\to 0$ corresponds to the global asymptotic stability of (9.1), and hence we see that (9.1) is *globally asymptotically stable if and only if* $r(A)<1$. In this case we shall say $A$ is *stable*. For computational criteria that the eigenvalues of a matrix lie in the unit circle, see [48].

**9.4.** *Exercise.* Show that

$$w_1(n)=\lambda_1^n,$$

$$w_j(n+1)=\sum_{k=0}^{n}\lambda_j^{n-k}w_{j-1}(k),\qquad j=2,\cdots,m.$$

**9.5.** *Exercise.* If the eigenvalues of $A$ are distinct, show that

$$w_1(n)=\lambda_1^n,$$

$$w_j(n)=\sum_{i=1}^{j}c_{ij}\lambda_i^n,$$

where

$$c_{ij}^{-1}=\prod_{1\leqq k<i\leqq j}(\lambda_i-\lambda_k).$$

**9.6.** *Exercise* (*variation of constants formula*). Show that the solution of the initial value problem ($f: J_+ \to R^m$)

$$x' = Ax + f(n), \quad x(0) = x^0$$

is at time $n+1$

$$x(n+1) = A^{n+1}x^0 + \sum_{k=0}^{n} A^{n-k}f(k).$$

**9.7.** *Exercise.* Show that the solution of

$$y(n+m) + a_1 y(n+m-1) + \cdots + a_m y(n) = g(n) \tag{9.5}$$

satisfying $y(0) = y(1) = \cdots = y(m-1) = 0$ is

$$y(n+1) = \sum_{k=0}^{n} w(n-k)g(k),$$

where $w$ is the $m$th principal solution of the homogeneous equation ($g(n) \equiv 0$); that is, $w$ is the solution of

$$y(n+m) + a_1 y(n+m-1) + \cdots + a_m y(n) = 0$$

satisfying

$$y(0) = \cdots = y(m-2) = 0, \quad y(m-1) = 1.$$

Since $A^{k+1} - A^k = A^k(A - I)$, we see that

$$A^{n+1} - I = (A - I) \sum_{k=0}^{n} A^k.$$

If $A^{n+1} \to 0$ ($r(A) < 1$), we obtain

$$(I - A)^{-1} = \sum_{k=0}^{\infty} A^k. \tag{9.6}$$

$R(\lambda) = (\lambda I - A)^{-1}$ is called the *resolvent* of $A$. We see, therefore, that if $|\lambda| > r(A)$,

$$R(\lambda) = (\lambda I - A)^{-1} = \sum_{k=0}^{\infty} \lambda^{-(k+1)} A^k. \tag{9.7}$$

It is also of interest to know when it is true that each solution of (9.1) approaches a point—which, of course, must be an equilibrium point. This is equivalent to $A^n \to B$ as $n \to \infty$. We already know that $A^n \to 0$ as $n \to \infty$ if $r(A) < 1$ and that $A^n$ is unbounded if $r(A) > 1$. The question is answered in the next exercise since the only case left is $r(A) = 1$. Note that $\lambda = 1$ a simple pole of the resolvent of $A$ is equivalent to 1 is a simple root of the minimal polynomial of $A$.

**9.8.** *Exercise.* Show that: If $r(A) = 1$, then $A^n$ converges if and only if $\lambda = 1$ is a simple pole of the resolvent of $A$ and is the only eigenvalue of $A$ on the unit circle. (This can be seen from the algorithm from computing $A^n$ or from the Jordan canonical form for $A$. See also [95, Chap. 1].)

**9.9.** *Exercise.* Show that: If $A^n \to B$ as $n \to \infty$, then $AB = BA = B$ and $B^2 = B$.

**9.10.** *Exercise.* When is it true that each solution of (9.1) is bounded?

To illustrate further the application of Liapunov's direct method and the use of Liapunov functions, we shall study a bit more the question of the stability of $A$.

The next criterion is the analogue of the one given originally by Liapunov for the real parts of all the eigenvalues of $A$ to be negative ($e^{At} \to 0$ as $t \to \infty$).

Let $V(x) = x^T Bx$, where $B$ is positive definite. Then, with respect to (9.1), $\dot{V}(x) = x^T(A^T BA - B)x$. Hence, if $A^T BA - B$ is negative definite, (9.1) is asymptotically stable by Exercise 7.11, and $A$ is stable. Conversely, suppose that $A$ is stable, and consider the equation

$$A^T BA - B = -C. \tag{9.8}$$

If it has a solution, then

$$-\sum_{k=0}^{n} (A^T)^k CA^k = (A^T)^{n+1})BA^{n+1} - B.$$

Letting $n \to \infty$, we see that the solution must be

$$B = \sum_{k=0}^{n} (A^T)^k CA^k.$$

It is easily verified that this is a solution and that, if $C$ is positive definite, it is positive definite.

Hence, we have shown that the following holds.

**9.11.** THEOREM. *If there are positive definite matrices B and C satisfying* (9.8), *then A is stable. Conversely, if A is stable, then given C,* (9.8) *has a unique solution B. If C is positive definite, B is positive definite.*

This result plays an important role in the theory of linear discrete control systems. It is also a converse theorem; if (9.1) is asymptotically stable, there is a positive definite quadratic Liapunov function $V$ with $-\dot{V}$ positive definite.

We now obtain an immediate consequence of Theorem 9.11 that will be useful to us in a moment. A symmetric real matrix $B$ is said to be *indefinite* if it is symmetric and if $x^T Bx$ takes on both positive and negative values ($B$ has both positive and negative eigenvalues).

**9.12.** COROLLARY. *If $r(A) > 1$ and* (9.8) *has a unique solution B for a positive definite matrix C, then B is symmetric and $x^T Bx$ takes on negative values.*

*Proof.* If $B$ is a solution, then so is $B^T$, and hence $B^T = B$. Since $r(A) > 1$, we know by Theorem 9.11 that $B$ is not positive definite. Also $B$ cannot be semipositive definite since, if it were, there would be an $x \neq 0$ with $Bx = 0$. But then $-x^T Cx = x^T A^T BAx \geqq 0$, a contradiction. ■

The question of when (9.8) has a unique solution is answered in the next exercise. This is the analogue of the result that $A_1X - A_2X = C$ has a unique solution if and only if $A_1$ and $A_2$ have no common eigenvalues.

**9.13.** *Exercise.* Let $A_1$ be an $m \times m$ matrix and $A_2$ an $n \times n$ matrix; $C$ and $X$ are $m \times n$ matrices. Show that: the equation $A_1XA_2 - X = C$ has a unique solution $X$ if and only if no eigenvalue of $A_1$ is a reciprocal of an eigenvalue of $A_2$. (Suggestion: If $A_1$ and $A_2$ satisfy the above condition, use the Hamilton–Cayley theorem to show that there is a polynomial $\phi(\lambda)$ of degree $k$ such that $\phi(A_1) = I$ and $\phi^*(A_2) = 0$, where $\phi^*(\lambda) = \lambda^k\phi(1/\lambda)$, and use this to show the sufficiency of the condition, since $X = A_1XA_2$ implies $\phi(A_1)X = A_1^kX\phi^*(A_2)$.)

In 1929 Perron in [91] investigated the question of when the stability of the linear approximation $x' = Ax$ determines the stability of the nonlinear equation

$$x' = Ax + f(n, x), \tag{9.9}$$

where $f(n, x)$ is $o(x)$ uniformly with respect to $n \geqq 0$ (i.e., given $\varepsilon > 0$ there is a $\delta > 0$ such that $\|x\| < \delta$ implies $\|f(n, x)\| < \varepsilon\|x\|$ for all $n \geqq 0$ and all $\|x\| < \delta$). Near the origin the effect of the nonlinearity should be negligible, and the stability should be determined by the linear approximation except in the critical case when $r(A) = 1$. This is the content of the next result. Stability here means stability of the equilibrium point at the origin. Since we have not discussed stability for nonautonomous systems we confine ourselves to

$$\dot{x} = Ax + f(x). \tag{9.10}$$

The result and proof is exactly the same for (9.11) and the asymptotic stability is uniform. Perron gave a different proof.

**9.14.** THEOREM (stability by the linear approximation). *Let $f(x)$ be $o(x)$. If $A$ is stable, then the origin is an asymptotically stable equilibrium point of* (9.10). *If $r(A) > 1$, the origin is unstable.*

*Proof.* Assume that $A$ is stable. Then there is a positive definite matrix $B$ satisfying (9.8). For convenience, take $C = I$, and let $V(x) = x^T Bx$. Then relative to (9.10)

$$\dot{V}(x) = -x^T x + 2x^T Bf(x) + V(f(x)).$$

For any $0 < \alpha < 1$ we can select $\delta$ sufficiently small that

$$\dot{V}(x) \leqq -\alpha x^T x \quad \text{for all} \quad \|x\| < \delta.$$

Hence $V$ and $-\dot{V}$ are positive definite, and the origin is asymptotically stable.

If $r(A) > 1$ select $\beta > 0$ so that no eigenvalue of $(1+\beta)^{-1/2}A$ is the reciprocal of an eigenvalue of $A$ and sufficiently small that $(1+\beta)^{1/2} < r(A)$. Then by Exercise 9.13 and Corollary 9.12 there is a matrix $B$ that is either negative definite or indefinite satisfying

$$A^T BA - B = \beta B - I.$$

Taking $V(x) = -x^T Bx$, we have again, for any $0 < \alpha < 1$ and $\delta$ sufficiently small,

$$\dot{V}(x) = \beta V(x) + W(x),$$

where $W(x) \geqq \alpha x^T x$ for all $\|x\| < \delta$. By Corollary 7.16 the origin is unstable. ■

**9.15.** *Exercise* (*an open question*). If $r(A) > 1$, the origin for the linear approximation is strongly unstable (is neither stable nor an attractor). Under what conditions on $A$ is the origin strongly unstable for the nonlinear equation (9.10)? One such condition is that all the eigenvalues of $A$ lie outside the unit of the circle.

Determining stability by the linear approximation was, and still is, important in many applications. Up until about 1950 it was almost the only mathematical approach to the design and analysis of control systems and feedback devices, and this goes back to Maxwell [69], [70], in 1868. It should, however, be kept in mind that the results are local and from a practical point of view may be completely misleading. An equilibrium can be asymptotically stable but yet its region of asymptotic stability can be so small that, as a practical matter, it is unstable. Conversely, it could be mathematically unstable, caused by a stable oscillation about the equilibrium, but yet the oscillation could be so small that its effect is negligible. The advantage of Liapunov's method, when it can be successfully applied, is that it takes into account the nonlinearities and yields information about the extent of the stability.

Nonnegative matrices $A = (a_{ij})$ are those for which $a_{ij} \geqq 0$, and we designate them by $A \geqq 0$. They arise and are important in many applications and have been studied for a long time (see [11], [28]). For instance, the linear difference equation $x' = Ax$ may be a mathematical model for a system where the state variables have meaning only when they are nonnegative ($x \geqq 0$)—populations, prices, number of particles, etc. Then $A$ will be nonnegative, since this is equivalent to $R^m_+$ positively invariant. There are many characterizations of stable nonnegative matrices, among which, for purposes of illustrating our results on stability and for later reference, we list a few in the following theorem.

**9.16.** THEOREM. *If $A \geqq 0$, the following are equivalent*:

(i) $r(A) < 1$,
(ii) $(A - I)^{-1} \leqq 0$,
(iii) $Ac < c$ *for some* $c > 0$,
(iv) $c^T A < c^T$ *for some* $c > 0$,
(v) $\mathcal{R}l(\sigma(A - I)) < 0$,
(vi) *The principal minors of* $I - A$ *are all positive.*

*Proof.* We shall show the equivalence of the first four. It is known (see [28]) that (iii), (v) and (vi) are equivalent. In Examples 8.4 and 8.5 we showed that (ii)$\Rightarrow$(i) and (iii)$\Rightarrow$(i). By (9.6) we see that (i)$\Rightarrow$(ii). To see that (ii)$\Rightarrow$(iii) take $c = (I - A)^{-1} b$ for any $b > 0$. Then, since $(I - A)^{-1}$ is nonsingular and nonnegative, $c > 0$. This proves the equivalence of the first three. The equivalence with (iv) follows upon noting that $r(A) = r(A^T)$. ■

**9.17.** *Exercise.* Show that: $r(A) \geqq r(|A|)$.

**10. Global asymptotic stability.** Let us note first the following result on global asymptotic stability that is an immediate consequence of Theorem 7.9.

**10.1.** COROLLARY. *If* (i) *$V$ is a Liapunov function of $\dot{x} = Tx$ on $R^m$,* (ii) *$G_c = \{x;\ V(x) < c\}$ is bounded for each $c$, and* (iii) *$M$ is compact, then $M$ is a global attraction. If, in addition,* (iv) *$V$ is constant on $M$, then $M$ is globally asymptotically stable.*

**10.2.** *Exercise.* Show that: If $V(x) \to \infty$ as $\|x\| \to \infty$, then $G_c$ is bounded for each $c$.

We want to raise and answer some questions concerning the global asymptotic stability of equilibrium points. We shall raise more questions than we answer. Consider

$$(10.1) \qquad x' = Tx, \qquad T(0) = 0.$$

We have placed the equilibrium point at the origin, and in place of saying the origin for (10.1) is globally asymptotically stable we shall say that (10.1) is globally asymptotically stable.

When $m = 1$, we have $\dot{x} = T(x) = a(x)x$, where $a(x) = T(x)/x$ for $x \neq 0$. A sufficient condition for global asymptotic stability is $|a(x)| < 1$ for $x \neq 0$. If $a(x) \geqq 0$, this condition is both necessary and sufficient. If $T(x)$ is $C^1$, then $a(x) = \int_0^1 T'(sx)\,ds$ ($T'$ is the derivative), and $|T'(x)| < 1$ for $x \neq 0$ is a sufficient condition for global asymptotic stability.

The general question is: *how do these conditions generalize for higher dimensions*? If $T$ is $C^1$, then one might try to find conditions on the Jacobian matrix $T' = (\partial T_i / \partial x_j)$ that imply global asymptotic stability. Or we might consider $T(x)$ in the form $T(x) = A(x)x$, where $A(x)$ (as always, we assume at least that $T$ is continuous on $R^m$) is an $m \times m$ matrix function, and study equations of the form

$$(10.2) \qquad x' = A(x)x.$$

If $T(0) = 0$ and $T$ is $C^1$, then $T(x) = \int_0^1 T'(sx)x\,ds$ so that $A(x) = \int_0^1 T'(sx)\,ds$ is always one $A(x)$ for which $T(x) = A(x)x$. But, in general, $B(x)x = 0$ for all $x$ does not imply $B(x) = 0$ for all $x$, so that the $A(x)$ of equation (10.2) need not be unique. Note, for instance, if $A(x)$ is continuous at the origin, then $A(0)$ stable implies asymptotic stability of the origin. This follows from Theorem 9.14. Similarly, if $T$ is $C^1$ and $T'(0)$ is stable.

**10.3. Notation.** For $B = (b_{ij})$, $|B| = (|b_{ij}|)$, $\|B\| = \max_{\|x\|=1} \|Bx\|$.

**10.4.** *Exercise.* For $\phi(B) = |B|$ or $\phi(B) = \|B\|$ show that:

(a) $\phi(B)$ is positive definite (i.e., $\phi(B) \geqq 0$ and $\phi(B) = 0$ implies $B = 0$).

(b) If $AB$ is defined, $\phi(AB) \leqq \phi(A)\phi(B)$.

(c) If $A+B$ is defined, $\phi(A+B) \leqq \phi(A)+\phi(B)$.
(d) $\phi(\alpha B)=|\alpha|\phi(B)$.

Possible generalizations of the sufficient conditions for global asymptotic stability for $m=1$ are: (for all $x \neq 0$)

$C_1$. $r(A(x))<1$.
$C_2$. $r(T'(x))<1$.
$C_3$. $r(|A(x)|)<1$.
$C_4$. $r(|T'(x)|)<1$.

Very little is known about these conditions, and there are many open questions. On the other hand, it is not difficult to show that each of the following conditions, which are much stronger than the above conditions and are by no means independent, is sufficient for global asymptotic stability: (for all $x \neq 0$)

$C_5$. $\|QT(x)\|<\|Qx\|$ for some nonsingular $Q$.
$C_6$. $|QT(x)|<|Qx|$ for some nonsingular matrix $Q$.
$C_7$. $|T(x)| \leqq B|x|$, $B$ stable.
$C_8$. $A^T(x)BA(x)-B$ is negative definite for some positive definite matrix $B$.
$C_9$. $T$ is $C^1$ and $T'^T(x)BT'(y)-B$ is negative definite, all $x \neq 0$, all $y \neq 0$ and some positive definite matrix $B$.
$C_{10}$. $|A(x)| \leqq B$ and $B$ stable.
$C_{11}$. $T$ is $C^1$ and $|T'(x)| \leqq B$ and $B$ stable.

**10.5.** *Exercise.* Show that:
(a) $C_5$ is equivalent to $C_8$;
(b) $C_9$ implies $C_8$;
(c) $C_{11}$ implies $C_{10}$ which implies $C_7$;
(d) $C_8$ implies $C_1$;
(e) $C_9$ implies $C_2$.

**10.6.** DEFINITION. Let $H$ be a closed set in $R^m$ and let $G$ be a positively invariant set in $R^m$ with $H \subset \bar{G}$. *Stability of H relative to G* means stability of $H$ with respect to the relative topology on $G$. Similarly, for the definition of $H$ is an *attractor relative to G*. If $H$ is stable and an attractor relative to $G$, then $H$ is *asymptotically stable relative to G*. If $H$ is *stable relative to G* and, for each $x \in G$, $T^n x \to H$, then $H$ is *globally asymptotically stable relative to G*.

Thus, Theorem 7.9 without the assumption that $M \subset G$ gives a sufficient condition for global asymptotic stability of $M$ relative to $G$. Another modification of Theorem 7.9 that is often useful is the following result.

**10.7.** THEOREM. *Let G be a bounded open set, positively invariant with respect to* (10.1). *Assume that V is a Liapunov function of* (10.1) *on G and that* $\dot{V}(x)<0$ *for* $x \in G$, $x \neq 0$. *Let* $E_0$ *be the set of points on the boundary of G where* $\dot{V}$ *vanishes, and let* $M_0$ *be the largest invariant set in* $E_0$. *Then all solutions starting in G approach* $M=M_0 \cup \{0\}$. (i) *If* $M=\{0\}$, *the origin is globally asymptotically stable relative to G.* (ii) *If* $0 \in G$, *then each solution starting in G approaches either* 0 *or* $M_0$ *as* $n \to \infty$. *If no solution can approach* $M_0$ *as* $n \to \infty$, *then the origin is asymptotically stable and is globally asymptotically stable relative to G.*

**10.8.** *Example. A discrete epidemic model (gonorrhea).* Consider two distinct (heterosexual) populations where infected members of one population can transmit a disease to a susceptible in the other population. Recovery is possible but there is no immunity, and the populations are constant. Let $x_i$ be the fraction of the population $P_i$ infected. Then $(1-x_i)$ is the fraction susceptible. A simple and not unreasonable discrete model is

$$x_1(n+1)=a_1x_2(n)(1-x_1(n))+(1-b_1)x_1(n),$$
$$x_2(n+1)=a_2x_2(n)(1-x_2(n))+(1-b_2x_2(n);$$

i.e.,

$$x'=A(x)x,$$

where $0<a_i<1$, $0<b_i<1$, and

$$A(x)=\begin{pmatrix} 1-b_1 & a_1(1-x_1) \\ a_2(1-x_2) & 1-b_2 \end{pmatrix}.$$

Let $G=\{x; 0<x_i<1\}$. Then, as is to be expected of the model, $G$ and $\bar{G}$ are positively invariant; in fact, $T(\bar{G}-\{0\})\subset G$. To see how this generalizes note that

$$A(x)=I+A_0+A_1(x),$$

where $A(x)\geqq 0$, $I+A_0\geqq 0$, and $A_1(x)\leqq 0$ for $x\in\bar{G}$. Also $A_1(0)=0$. For $n=2$,

$$A_0=\begin{pmatrix} -b_1 & a_1 \\ a_2 & -b_2 \end{pmatrix}$$

and

$$A_1(x)=\begin{pmatrix} 0 & -a_1x_1 \\ -a_2x_2 & 0 \end{pmatrix}.$$

There are two cases: (i) $r(I+A_0)\leqq 1$, and (ii) $r(I+A_0)>1$.

*Case* (i). By applying Theorem 9.16 to $(1-\varepsilon)A_0$ and letting $\varepsilon\to 0^+$ we see that $r(I+A_0)\leqq 1$ if and only if the principal minors of $-A_0$ are all nonnegative, which is equivalent to the existence of a $c>0$ with $c^\tau A_0\leqq 0$. Taking $V(x)=c^\tau x$, we have

$$\dot{V}(x)=c^\tau(A_0+A_1(x))x\leqq 0 \quad \text{for} \quad x\in\bar{G}$$

and $\dot{V}(x)<0$ for $x\in G$. However, since $T(\bar{G}-\{0\})\subset G$, we see that $M=\{0\}$ and by Theorem 10.7 the origin is globally asymptotically stable relative to $\bar{G}$.

For $n=2$, this case corresponds to $\det A_0=b_1b_2-a_1a_2\leqq 0$ and we can take $c^\tau=(b_2, a_1)$. Then

$$\dot{V}(x)=(a_1a_2-b_1b_2)x_1-a_1(a_2+b_2)x_1x_2.$$

At worst ($\det A_0=0$) $E_0$ is the intersection of the coordinate axes with $\bar{G}$, and $M=\{0\}$.

*Case* (ii). From the linear approximation, $x' = (I + A_0)x$, we know that the origin is unstable. Note that near the origin in $\bar{G}$ both $V(x)$ and $\dot{V}(x)$ are positive, and hence no solution starting in $\bar{G} - \{0\}$ can approach the origin as $n \to \infty$. (Actually, it is not difficult to see that all solutions must approach an invariant set in $G$ and that there must be a unique equilibrium point in $G$.)

For $n = 2$ this equilibrium point is

$$x_1^0 = \frac{1 - \gamma_1\gamma_2}{1 + \gamma_1} \quad \text{and} \quad x_2^0 = \frac{1 - \gamma_1\gamma_2}{1 + \gamma_2},$$

where $\gamma_i = b_i/a_i$, and $r(I + A_0) > 1$ corresponds to $b_1b_2 - a_1a_2 < 0$ (i.e., $\gamma_1\gamma_2 < 1$). Changing to coordinates $u = x - x^0$ about this equilibrium point, we have

$$u' = B(x)u,$$

where

$$B(x) = \begin{pmatrix} 1 - a_1\rho & a_1(1 - x_1) \\ a_2(1 - x_2) & 1 - a_2\rho^{-1} \end{pmatrix}$$

and $\rho = (1 + \gamma_1)/(1 + \gamma_2)$. Noting that

$$(a_2, a_1\rho)\begin{pmatrix} -a_1\rho & a_1 \\ a_2 & -a_2\rho^{-1} \end{pmatrix} = 0,$$

we take

$$V(u) = a_2|u_1| + a_1\rho|u_2|.$$

Then, if $1 - a_1\rho \geqq 0$ and $1 - a_2\rho^{-1} \geqq 0$,

$$\dot{V}(u) \leqq -a_1a_2(x_1|u_2| + \rho x_2|u_1|) \leqq 0$$

for $x \in G$. In the $x$-coordinates the set $M_0$ of Theorem 10.7 is the origin and, since no solution starting in $\bar{G} - \{0\}$ can approach $M_0$, $x^0(u = 0)$ is globally asymptotically stable relative to $\bar{G} - \{0\}$. Note, for instance, that $1 - a_1\rho \geqq 0$ and $1 - a_2\rho^{-1} \geqq 0$ if $a_i + b_i \leqq 1$. If the unit of time is sufficiently small, this condition is always satisfied (as the unit of time goes to zero, $a_i + b_i \to 0$). One might suspect for $\gamma_1\gamma_2 < 1$ that $x^0$ is always globally asymptotically stable relative to $G - \{0\}$ but our method depends upon $B(x) \geqq 0$ on $G$. It is true (Exercise 10.9) that $x^0$ is always asymptotically stable.

**10.9.** *Exercise.* In the above example show, by examining the linear approximation about $u = 0$, that $x^0$ is asymptotically stable when $\gamma_1\gamma_2 < 1$. (For dimension 2 the characteristic roots of a matrix $B$ lie inside the unit circle if and only if $|\det B| < 1$ and $|\text{trace } B| < 1 + \det B$. For general criteria and computational algorithms, see [48].)

**10.10.** *Exercise.* In the epidemic model above introduce a third population group that infects members of the group as well as those outside (in the language of the street they are AC-DC), and discuss the model.

**11. Stability under perturbations.** So far, we have been concerned only with what happens to the asymptotic behavior of $T^n x$ under perturbations of the initial state $x$. For applications, the more important question is: what happens to stability under perturbations of $T$? It was not until the 1930's, after converse theorems on the existence of Liapunov functions had been studied, that even partial answers to this question were given for ordinary differential equations (see [39], [50]). The analogous theory for difference equations was given by Halanay in [34] in 1963. The converse theorems are too important a theoretical tool not at least to mention, which we do without proofs. Following the work of Auslander and Seibert in [9], [10], [96] for abstract continuous dynamical systems, we state a result stronger than Halanay's on stability under perturbations of $T$, which is—with slightly stronger assumption on the continuity of $T$—that asymptotic stability is equivalent to a strong stability under perturbations. This is the practical significance of asymptotic stability.

Consider

$$x' = T(x), \quad T(0) = 0. \tag{11.1}$$

Now our definitions of stability and asymptotic stability, which are Liapunov's, are in terms of perturbations of initial conditions. In the real world, where nothing is known exactly, the system is constantly perturbed and a better model for the perturbations is

$$x' = T(x) + P(n, x), \tag{11.2}$$

where $P(n, x)$ is unknown but hopefully not too large. Certainly, simple stability of (11.1) is too fragile to expect it to imply a stability of the perturbed system (11.2). To establish a relationship between asymptotic stability and stability of the perturbed system, it seems to be necessary to assume that $T$ is Lipschitz continuous near the equilibrium state; that is, for some $L_1 > 0$ and some $r > 0$,

$$\|T(x) - T(y)\| \leqq L_1 \|x - y\| \quad \text{for all} \quad \|x\| < r \quad \text{and} \quad \|y\| < r. \tag{11.3}$$

The proof uses the following converse theorem [34].

**11.1 Theorem.** *If $T$ is Lipschitz continuous near the origin and the origin is asymptotically stable, there exists a positive definite Liapunov function $V$ which is Lipschitz continuous near the origin with $\dot{V}$ negative definite; i.e., for some $r_0 > 0$ and some $L_2 > 0$, $|V(x) - V(y)| \leqq L_2 \|x - y\|$ for all $\|x\| < r_0$ and $\|y\| < r_0$; $V(0) = 0$, and $V(x) > 0$ and $\dot{V}(x) < 0$ for all $0 < \|x\| < r_0$.*

**11.2. Definition.** Let $x^*(n, x)$ denote the solution of (11.2) satisfying $x^*(0, x) = x$. The origin, an equilibrium state of (11.1), is said to be *stable under perturbations* if for some $\varepsilon_0$ and each $\varepsilon > 0$ there exist $\delta_1(\varepsilon)$ and $\delta_2(\varepsilon)$ such that $\|x\| < \delta_1(\varepsilon)$ and $\|P(n, y)\| \leqq \delta_2(\varepsilon)$ for all $n \in J_+$ and all $\|y\| \leqq \varepsilon_0$ imply $\|x^*(n, x)\| < \varepsilon$ for all $n \in J^+$. If, in addition, there is a $\delta_0$, an $r_0$, and an $N(\varepsilon)$ such that $\|x\| < \delta_0$ and

$\|P(n, y)\| \leqq \delta_2(\varepsilon)$ for all $n \in J_+$ and all $\|y\| \leqq r_0$ imply $\|x^*(n, x)\| < \varepsilon$ for all $n \geqq N(\varepsilon)$, the origin is said to be *strongly stable under perturbations*.

It can be shown using Theorem 11.1 that the following holds.

**11.3.** THEOREM. *If $T$ is Lipschitz continuous in a neighborhood of the origin, then the origin is strongly stable under perturbations if and only if for the unperturbed system* (1.11) *it is asymptotically stable.*

In [34], under the same conditions on $T$, Halanay showed only that asymptotic stability of the unperturbed system implies that the origin is stable under perturbations.

# CHAPTER 2

## Ordinary Differential Equations. Local Dynamical Systems

**1. Introduction.** Stability theory was developed first for systems of ordinary differential equations beginning with Liapunov [66] in 1892. The concepts of positive and negative limit sets, which play a central role in our extension of Liapunov's theory, go back at least to Hadamard in 1897. In [31] he called a limit set "le domaine du mouvement" and showed that it was invariant. In 1912 in [14] Birkhoff studied limit sets in more detail and, among other things, pointed out that a solution approaches its positive limit set as $t \to \infty$. In this paper and in Chapter VII of his book [15], first published in 1924, Birkhoff laid the foundations for a general theory of dynamical systems (see [29], [84], [99], [112], [13], [59]).

Having already presented stability theory for difference equations, we now follow the same road for differential equations. The changes this necessitates are fairly obvious, and we are able to proceed rapidly. In any case the proofs are easy and are almost identical to those in [58]. In [58], however, solutions were not assumed to be unique, and, in order that $\dot{V}$ be well-defined, $V$ was assumed to be a bit smoother (locally Lipschitzian). Also in [58] the closure of the set $G$ was assumed to be in the domain of definition $G^*$ of the differential equation. This is not necessary—and is awkward for some applications—and here we drop that assumption. For ordinary differential equations, solutions are unique in both directions of time, the motions are continuous curves, and solutions can go to infinity in finite time. Thus, although the road is the same as in Chapter 1, the scenery is different and in many respects more interesting.

**2. Autonomous ordinary differential equations.**

**2.1. Notation.**

$R = (-\infty, \infty)$, the real numbers;

$R_+ = [0, \infty)$ and $R_- = (-\infty, 0]$;

$f: G^* \to R^n$, where $G^*$ is an open set in $R^n$.

We assume throughout that $f$ is continuous and that the solution $\pi(t, x^0)$ in $G^*$ of

$$\frac{dx}{dt} = \dot{x} = f(x) \tag{2.1}$$

satisfying $\pi(0, x^0)$ is unique.

**2.2.** DEFINITION. Let $\phi: (\alpha, \omega) \to G^*$, where $-\infty \leqq \alpha < 0 < \omega \leqq \infty$. A point $p$ is said to be a *positive* (*negative*) *limit point* of $\phi$ if there is a sequence $t_n \in (\alpha, \omega)$ such that $t_n \to \omega$ ($t_n \to \alpha$) and $\phi(t_n) \to p$ as $n \to \infty$. The set $\Omega(\phi)$ ($A(\phi)$) of all positive (negative) limit points of $\phi$ is called the *positive* (*negative*) *limit set* of $\phi$. The interval $(\alpha, \omega)$ is said to be *maximal* (this is relative to $G^*$) if $\omega < \infty$ implies $\Omega(\phi) \cap G^*$ is empty and if $-\infty < \alpha$ implies $A(\phi) \cap G^*$ is empty.

**3. Basic properties of solutions.** From the theory of ordinary differential equations we have the following properties of solutions (basic to our needs):

$P_1$ (*Existence*). Each solution $\pi(t, x)$ of (2.1) satisfying $\pi(0, x) = x$ has for each $x \in G^*$ a maximal interval of definition $I(x) = (\alpha(x), \omega(x))$, $-\infty \leqq \alpha < 0 < \omega \leqq \infty$.

$P_2$ (*Semigroup property*; *uniqueness*). For all $s \in I(x)$ and all $t \in I(\pi(s, x))$, $t + s \in I(x)$ and $\pi(t, \pi(s, x)) = \pi(t + s, x)$.

$P_3$ (*Continuous dependence*). $\pi$ is continuous; i.e., $(t_n, x^n) \in I(x^n) \times G^*$ and $(t_n, x^n) \to (t, x) \in I(x) \times G^*$, then $\pi(t_n, x^n) \to \pi(t, x)$.

$P_4$. $I(x)$ is lower semicontinuous on $G^*$; i.e., if $x^n \to x \in G$, then $I(x) \subset \liminf I(x^n) = \bigcup_{k=1}^{\infty} \bigcap_{n=k}^{\infty} I(x^n)$ (which means, if $t \in I(x)$, then $t \in I(x^n)$ for all $n$ sufficiently large).

The continuity property that we need here is; given $x^n \to x \in G^*$ and $t \in I(x)$, then $\pi(t, x^n) \to \pi(t, x)$. This weaker property is implied by $P_3$ and $P_4$. *Local dynamical systems* are mappings $\pi$ with the above properties. For a *dynamical system*, $I(x) = (-\infty, \infty)$ for all $x \in G^*$.

**3.1.** DEFINITION. A solution $\pi(t, x)$ is said to be *positively* (*negatively*) *precompact* if it is bounded for all $t \in [0, \omega(x))$ ($t \in (\alpha(x), 0]$) and has no positive (negative) limit points on the boundary of $G^*$. This is precompactness relative to $G^*$.

Positive precompactness corresponds to the older terminology "positively Lagrange stable". Note, of course, since $\omega(x)$ is maximal, that $\pi(t, x)$ positively precompact implies $\omega(x) = \infty$. We use $A(x)$ and $\Omega(x)$ to denote the positive and negative limit sets of $\pi(t, x)$.

**3.2.** *Exercise*. Show that:

$$\Omega(x) = \bigcap_{0 < \beta < \omega(x)} \overline{\pi([\beta, \omega(x)), x)}.$$

**4. Invariance.**

**4.1.** DEFINITION. Relative to (2.1), a set $H \subset R^n$ is said to be *positively* (*negatively*) *invariant* if $x \in H \cap G^*$ implies $\pi(t, x) \in H$ for all $t \in [0, \omega(x))$ ($t \in [\alpha(x), 0]$). $H$ is said to be *weakly invariant* if it is positively and negatively invariant. If, in addition, $I(x) = (-\infty, \infty)$ for each $x \in H \cap G^*$, $H$ is said to be *invariant*.

Note, of course, that, if $H$ is precompact relative to $G^*$ and weakly invariant, then it is invariant.

**4.2.** *Exercise.* Show that:

(a) The closure of a positive (negatively) invariant set is positive (negatively invariant), and hence the closure of a weakly invariant set is weakly invariant.

(b) If $H$ is invariant and precompact relative to $G^*$, then $\bar{H}$ is invariant.

**4.3.** *Exercise.* Let $E$ be a given set in $G^*$, and let $M$ be the largest (by inclusion) invariant set in $E$. Show that:

(a) $M$ is the union of all motions defined on $(-\infty, \infty)$ that remain in $E$ for all $t$.

(b) $x$ is in $M$ if and only if $I(x)=(-\infty, \infty)$ and $\pi(t, x)$ is in $E$ for all $t$.

(c) If $E$ is compact, then $M$ is compact.

Compare with Exercise 4.10 in Chapter 1.

**5. Basic properties of limit sets.** We consider here only positive limit sets. The results for negative limit sets are the same with $t$ replaced by $-t$.

**5.1.** THEOREM. *Every positive limit set is closed and weakly invariant.*

*Proof.* It follows either directly or from Exercise 3.2 that each positive limit set $\Omega(x)$ is closed. Let $y \in \Omega(x) \cap G^*$, and let $t \in I(y)$. Since $\Omega(x) \cap G^*$ is nonempty, $\omega(x)=\infty$, and there is a sequence $t_n$ such that $t_n \to \infty$ and $\pi(t_n, x) \to y$. As noted below $P_4$ above, for all $n$ sufficiently large, $t \in I(\pi(t_n, x))$ and $\pi(t, \pi(t_n, x))= \pi(t+t_n, x) \to \pi(t, y)$ as $n \to \infty$. Therefore $\pi(t, y) \in \Omega(x)$, and $\Omega(x)$ is weakly invariant. ■

**5.2.** THEOREM. *If $\pi(t, x)$ is positively precompact, then $\Omega(x)$ is in $G^*$, and is nonempty, compact, connected, invariant, and is the smallest closed set that $\pi(t, x)$ approaches as $t \to \infty$.*

*Proof.* The positively precompactness of $\pi(t, x)$ implies that $\Omega(x)$ is nonempty, in $G^*$, and compact. Since, by the theorem above, $\Omega(x)$ is weakly invariant, it is invariant. The remainder of the proof is similar to that of Theorem 5.2 of Chapter 1 (connectedness is even easier). ■

An example of a positive limit set that is not invariant and not connected is given in Fig. 2 of Appendix A. Note also in this example that solutions do not approach their positive limit set.

**5.3.** *Exercise.* Show that: If $\Omega(x) \subset G^*$ is nonempty and compact, then $\pi(t, x) \to \Omega(x)$ as $n \to \infty$. Hence conclude, if $\Omega(x) \subset G^*$ is nonempty, that $\Omega(x)$ is compact if and only if $\pi(t, x)$ is positively precompact.

**6. Liapunov functions. An extension of Liapunov's direct method.** Let $V: G^* \to R$ and define

$$\dot{V}(x)=\liminf_{t \to 0^+} \frac{1}{t}(V(\pi(t, x)-V(x)). \tag{6.1}$$

Thus, if $\phi(t)=V(\pi(t, x))$ and $D_+\phi(t)$ is the right lower derivative, then $D_+\phi(t)= \dot{V}(\pi(t, x))$. There is some reason for defining $\dot{V}$ in this generality, although its computation would seem to depend on knowing the solutions $\pi(t, x)$. If uniqueness of solutions were not being assumed, then it would not be clear that $\dot{V}(x)$ is well-defined. (In [58] and Appendix A uniqueness is not assumed.) In applications, $V$ is usually $C^1$ and $\dot{V}(x) = V'(x) \cdot f(x)$, where $V'(x) = \text{grad } V(x)$, and $\dot{V}$ can be computed directly from the differential equation.

**6.1.** DEFINITION. Let $V: G^* \to R$, and let $G$ be any subset of $G^*$. $V$ is said to be a *Liapunov function of* (2.1) *on* $G$ if (i) $V$ is continuous, and (ii) $\dot{V}(x) \leqq 0$ for all $x \in G$.

The following is a standard result in integration theory (see, for example, [71, p. 200]).

**6.2.** LEMMA. *Let* $\phi: [a, b] \to R$. *If*

(i) $\phi$ *is left lower semicontinuous on* $[a, b)$ (*i.e.*, $\liminf_{t \to t_0^-} \phi(t) \geqq \phi(t_0)$ *for each* $a \leqq t_0 < b$),

(ii) $D_+\phi(t) = \liminf_{h \to 0^+} (1/t)(\phi(t+h) - \phi(t)) \leqq 0$ *for all* $t \in [a, b)$ *except at most a countable number of points*,

*then* $\phi(t)$ *is nonincreasing on* $[a, b)$, *hence differentiable almost everywhere, and* $\phi(t) - \phi(a) \leqq \int_a^t \phi'(s)\, ds$ *for* $t \in [a, b)$.

Hence, we see that if $V$ is a Liapunov function of (2.1) on $G$, and $\pi(t, x)$ is in $G$ for $t \in [0, \beta]$, $\beta < \omega(x)$, then $V(\pi(t, x))$ is nonincreasing on $[0, \beta)$; in fact, $V(\pi(t, x)) - V(x) \leqq \int_0^t \dot{V}(\pi(s, x))\, ds$ on $[0, \beta)$.

**6.3. Notation.** Relative to a Liapunov function $V$ of (2.1) on $G$ we introduce the following notations:

$$E = \{x; \dot{V}(x) = 0, x \in \bar{G} \cap G^*\}, \tag{6.2}$$

$M$ is the largest invariant set in $E$, $M^*$ is the largest weakly invariant set in $E$, and $M^+$ is the largest positively invariant set in $E$. If $M^*$ is compact, then $M = M^*$. Note also that $M \subset M^* \subset M^+$. The set $M^+$ is the easiest to identify, and, although these sets can be distinct, it usually turns out that $M = M^+$.

**6.4.** THEOREM (invariance principle). *Let* $V$ *be a Liapunov function of* (2.1) *on* $G$, *and let* $x(t) = \pi(t, x^0)$ *be a solution of* (2.1) *that remains in* $G$ *for all* $t \in [0, \omega(x^0))$. *Then, for some* $c$, $\Omega(x^0) \cap G^* \subset M^* \cap V^{-1}(c)$. *If* $x(t)$ *is precompact, then* $x(t) \to M \cap V^{-1}(c)$.

*Proof.* Assume that $y \in \Omega(x^0) \cap G^*$. Then $\omega(x^0) = \infty$, and there is a sequence $t_n$ such that $x(t_n) \to y$ and $t_n \to \infty$ as $n \to \infty$. Now $V(x(t_n)) \to V(y)$ as $n \to \infty$; since $V(x(t))$ is nonincreasing with respect to $t$, $V(x(t)) \geqq V(y)$ for all $t \geqq 0$ and $V(x(t)) \to c$ as $t \to \infty$. Hence, for this $c$, $V(y) = c$ for each $y \in \Omega(x^0)$. Now, $\Omega(x^0)$ is weakly invariant, and therefore $\Omega(x^0) \subset E$, and hence in $M^*$. This proves the first conclusion. If $x(t)$ is precompact, then $\Omega(x^0)$ is invariant and $\Omega(x^0) \subset M \cap V^{-1}(c)$. Since $x(t) \to \Omega(x^0)$ as $t \to \infty$, $x(t) \to M \cap V^{-1}(c)$ as $t \to \infty$. ■

When $x(t) = \pi(t, x^0)$ is precompact, we know also that $\Omega(x^0)$ is connected. The following result is then an immediate consequence of Theorem 6.4, and shows that Liapunov functions can be used to establish the existence of equilibrium points (zeros of $f(x)$).

**6.5.** COROLLARY. *Let $V$ be a Liapunov function of* (2.1) *on $G$ and let $x(t)$ be a precompact solution of* (2.1) *that remains in $G$ for all $t \geqq 0$. If the points of intersection of $M$ (or $E$) with $V^{-1}(c)$ are isolated for each $c$, then $x(t)$ approaches an equilibrium point of* (2.1) *as $t \to \infty$.*

**6.6.** *Example.*

$$\dot{x}_1 = -x_1|x_2|^\alpha,$$
$$\dot{x}_2 = -x_2|x_1|^\beta, \qquad \alpha > 0, \beta > 0.$$

Let $V = x_1^2 + x_2^2$. Then $\dot{V}(x) = -2(x_1^2|x_1|^\alpha + |x_1|^\beta x_2^2)$, and $V$ is a Liapunov function on $R^2$. Here $E = M$ is the union of the coordinate axes, and, for each $c$, $V^{-1}(c) \cap M$ is the set of points $(0, \pm c)$, $(\pm c, 0)$. Each solution is bounded and therefore approaches one of these equilibrium points as $t \to \infty$.

Another corollary of Theorem 6.4, that plays a role in establishing instabilities, is the following.

**6.7.** COROLLARY. *Let $V$ be a Liapunov function of* (2.1) *on $G$, and let $x(t) = \pi(t, x^0)$ be a solution of* (2.1) *in $G$ for all $t \in [0, \omega(x^0))$. Assume $G^* = R^n$ (or that $\Omega(x^0) \subset G^*$). If $M^*$ is empty or if $x(t)$ has no positive limit points in $M^*$, then $x(t) \to \infty$ in finite time or as $t \to \infty$.*

**6.8.** *Example. Problem.* If $x(t)$ is a solution of $\ddot{x} + g(x, \dot{x}) = 0$, where $g(x, 0) < 0$ for $x > 0$, satisfying $x(0) > 0$ and $\dot{x}(0) > 0$, then $x^2(t) + \dot{x}^2(t) \to \infty$ in finite time or as $t \to \infty$.

*Solution.* An equivalent system is $\dot{x} = y$, $\dot{y} = -g(x, y)$. Take $V = -x$ and $G = \overline{R_+^2}$ and apply the above corollary (see [65], where this equation is discussed in detail; however, this instability result is not covered there).

Let us look a bit more at this example since it suggests Theorem 7.12 of the next section. The equation can also be written

$$\ddot{x} + f(x, \dot{x}) + g(x) = 0;$$

$g(x) = g(x, 0)$ and $f(x, \dot{x}) = g(x, \dot{x}) - g(x, 0)$. We have a conservative force $g(x)$ and damping $f(x, \dot{x})$. An equivalent system is $\dot{x} = y$, $\dot{y} = -g(x) - f(x, y)$. The total energy is $W = \frac{1}{2}y^2 + G(x)$, where $G(x) = \int_0^x g(s)\, ds$, and $\dot{W} = -yf(x, y)$. The above case corresponds to $g(x) < 0$ for $x > 0$ (a repulsive force for $x > 0$).

Assume that $yf(x, y) > 0$ for $y \neq 0$ (positive damping) and $xg(x) < 0$ for $x \neq 0$. Then $W$ is a Liapunov function on $R^2$, and the region $G$ defined by $W < 0$ is positively invariant and nonempty. Here $M = M^*$ is the origin, and, since no solution starting in $G$ can approach the origin, we see by the above corollary that each such solution approaches $\infty$ in finite time or as $t \to \infty$. No damping can stabilize this system.

**7. Stability and instability.** We shall restrict ourselves here to discussing the stability of sets $H$ in $G^*$ that are compact in $R^n$. Then the sets $B_\delta(H) = \{X; \rho(x, H) < \delta\}$ generate a complete neighborhood system of $H$ (for $\delta$ sufficiently small $B_\delta(H) \subset G^*$).

**7.1.** DEFINITION. A compact set $H \subset G^*$ is said to be *stable*, if given a neighborhood $U$ of $H$, there is a neighborhood $W$ of $H$ such that $x \in W$ implies $\pi(t, x) \in U$ for all $t \geqq 0$.

**7.2.** *Exercise.* Show that: If a compact set $H$ is stable, then it is positively invariant and, for each $x \in H$, $\omega(x) = \infty$. In particular, if a point $x$ of $G^*$ is stable, it is an equilibrium point.

**7.3. Notation.** For a set $H \subset G^*$ define $\hat{H}$ as follows: $z \in \hat{H}$ if there exists a sequence $x^n \in G^*$ and $t_n \in [0, \omega(x^n))$ such that $\pi(t_n, x^n) \to z$ as $n \to \infty$. $\hat{H}$ is called the *positive prolongation of H* (see [9], [10]).

Note, of course, that $H \subset \hat{H}$. In the example in Fig. 2 of the Appendix the solutions spiral out from an equilibrium point. The positive prolongation of that equilibrium point is the closed region bounded by the parallel lines $y = 0$ and $y = 1$. It is weakly invariant but not invariant.

**7.4.** *Exercise.* Show that:

(a) If $H$ is a compact positively invariant set in $G^*$, then $H$ is stable if and only if $\hat{H} = H$.

(b) If $H$ is a compact invariant set in $G^*$, then $\hat{H}$ is weakly invariant.

**7.5.** DEFINITION. A compact set $H \subset G^*$ is an *attractor* if there is a neighborhood $U$ of $H$ such that $x \in U$ implies $\pi(t, x) \to H$ as $t \to \infty$. If $\pi(t, x) \to H$ for each $x \in G^*$, $H$ is called a *global attractor.* If $H$ is both stable and an attractor, $H$ is said to be *asymptotically stable. H* is said to be *globally asymptotically stable* if it is stable and a global attractor.

**7.6.** DEFINITION. *Instability* means not stable. If $H$ is not stable and is not an attractor, $H$ is said to be *strongly unstable.*

**7.7.** *Exercise.* The region of attraction $\mathcal{R}(H)$ of a set $H$ in $G^*$ is the set of all $x \in G^*$ such that $\pi(t, x) \to H$ as $t \to \infty$. If a compact set $H \subset G^*$ is asymptotically stable, show that:

(a) $\mathcal{R}(H)$ is open,

(b) $\mathcal{R}(H)$, $\partial\mathcal{R}(H)$ and $\mathcal{C}\mathcal{R}(H)$ are weakly invariant.

The following type of result has proved to be useful and is a direct consequence of Exercise 7.4 and the invariance principle.

**7.8.** THEOREM. *Let G be a positively invariant open set in $G^*$ with the property that each solution starting in G is bounded and has no positively limit points on the* *boundary of G. If* (i) *V is a Liapunov function of* (2.1) *on G*, (ii) $M^0=\overline{M\cap G}\subset G$, *and* (iii) $M^0$ *is compact, then* $M^0$ *is an attractor and* $G\subset\mathcal{R}(M^0)$. *If, in addition,* (iv) *V is constant on the boundary of* $M^0$, *then* $M^0$ *is asymptotically stable.*

**7.9.** *Remark.* A result similar to this, but less general, is given in [58] (Corollary 1 and Theorem 3). There is, however, a mistake in the statement and proof of Corollary 1. Since $\dot{V}(x)$ need not be continuous, $E$ and $M$ may not be closed. Both Corollary 1 and Theorem 3 in [58] are correct if $M$ is replaced by $\bar{M}$. If $V$ is assumed to be $C^1$, then $M$ is closed, and the above assumption that $\overline{M\cap G}\subset G$ is equivalent to $M\cap G$ is closed. Here we allow the possibility that points of $M$ are on the boundary of $G$. The next example provides a reason for introducing this complication.

**7.10.** *Example.* $\ddot{x}+a\dot{x}+2bx+3x^2=0$, $a>0$ and $b>0$. There is equivalent to $\dot{x}=y$, $\dot{y}=-2bx-ay-3x^2$. There are equilibrium points at (0, 0) and $(-\frac{2}{3}b, 0)$. Let $V=\frac{1}{2}y^2+bx^2+x^3$, the total energy of the system. Then $\dot{V}=-ay^2$, and $V$ is a Liapunov function on $R^2$; $E$ is the $x$-axis and $M$ consists of the two equilibrium points. The region $V<\frac{4}{27}b^3$ is the union of two components $G_1$ and $G_2$. Let $G_1$ be the bounded component containing the origin (to the right of $(-\frac{2}{3}b, 0)$), and let $G_2$ be the unbounded component to the left of $(-\frac{2}{3}b, 0)$. Both $G_1$ and $G_2$ are positively invariant, and it is clear that no solution starting from inside either $G_1$ or $G_2$ can approach $(-\frac{2}{3}b, 0)$. The conditions of Theorem 7.8 are satisfied for $G_1$, and, since $M_1$ is the origin, the origin is asymptotically stable. $G_1$ is a measure of the stability of the origin for all $a>0$; $G_1$ is in the region of attraction to the origin. For $G_2$, $M_2$ is the unstable equilibrium point $(-\frac{2}{3}b, 0)$. Since no solution starting in $G_2$ can approach $(-\frac{2}{3}b, 0)$, each solution starting in $M_2$ approaches infinity in finite time or as $t\to\infty$.

**7.11.** *Exercise.* Consider the equation

$$\frac{d^3y}{dt^3}+b_1\frac{d^2y}{dt^2}+(1+b_2)\frac{dy}{dt}+b_1y+\frac{d}{dt}(y^3)=0,$$

$b_1>0$ *and* $b_2>0$. Show that this equation is equivalent to the system

$$\dot{x}=A(x)x,$$

where

$$A(x)=\begin{pmatrix}-b_1 & 1 & 0\\ -x_1^2-b_2 & 0 & 1\\ 0 & -1 & 0\end{pmatrix},$$

and conclude that the origin is globally asymptotically stable. (This equation was given by Singh in [100] as a counterexample to an intuitive method, called by engineers "harmonic linearization," for approximating and establishing the existence of periodic solutions of nearly nonlinear systems.)

We shall state one instability result, which generalizes Ĉetaev's instability theorem. The proof is a direct consequence of Theorem 6.4. The statement of the theorem is complicated by the fact that it includes the possibility that the equilibrium point may be either on the boundary or the interior of $U$.

**7.12.** THEOREM. *Let $y \in G^*$ be an equilibrium point of* (2.1) *contained in the closure of an open set $U \subset G^*$. Let $N$ be a neighborhood of $y$. Assume that*: (i) *$V$ is a Liapunov function of* (2.1) *on $G = U \cap N$, and $M \cap G$ is either empty or the point $y$,* (ii) *$V(y) = 0$ and $V(x) = 0$ on that part of the boundary of $G$ inside $N$, and* (iii) *$V(x) < 0$ for $x \in G$, $x \neq y$. Then $y$ is unstable. In fact, if $N_0 \subset G^*$ is any bounded neighborhood of $y$ properly contained in $N$, then each solution starting in $G_0 = N_0 \cap G$ at a point other than $y$ leaves $N_0$. If $U$ is positively invariant, then no solution starting in $U$ at a point other than $y$ can approach $y$ as $t \to \infty$, and $y$ is strongly unstable.*

**7.13.** *Example.* Consider

$$\dot{x} = xy + xy^2 + \cdots,$$

$$\dot{y} = -y + x^2 + \cdots,$$

where the dots indicate that higher order terms have been neglected. In the linear approximation the origin is stable but, as we shall see, the nonlinearities cause instability. Let $V = y^2 - x^2$. Then $\dot{V} = -2y^2(1 + x^2) + \cdots$. With $U$ the region defined by $V < 0$ and $N$ a sufficiently small neighborhood of the origin, the conditions of Theorem 7.12 are satisfied and the origin is unstable. If there were no higher order terms, we could conclude that each solution starting in $V < 0$ approaches infinity in finite time or as $t \to \infty$.

**8. Vector Liapunov functions.** Just as in § 8 of Chapter 1, the usual notion of a vector Liapunov function introduces nothing essentially new (see, for example, [60]). We turn, therefore, to the more general notion of a vector Liapunov function suggested by the Liapunov function (they do not call it that) used by Arrow, Block and Hurwicz in [4].

**8.1. Notation.**

$$w: G^* \to R^m,$$

$$W(x) = \max_i w_i(x),$$

$$J(x) = \{j;\ w_j(x) = W(x)\},$$

$$u^\tau = (1, 1, \cdots, 1),$$

$$\mathring{w}(x) = \liminf_{t \to 0^+} (1/t)(w(\pi(t, x)) - W(x)u).$$

Note that, if $w$ is continuous,

$$\mathring{w}_j(x) = \begin{cases} -\infty & \text{if } j \notin J(x), \\ \dot{w}_j(x) & \text{if } j \in J(x). \end{cases}$$

**8.2.** DEFINITION. $w$ is a *vector Liapunov function of* (2.1) *on* $G$ if (i) $w$ is $C^1$ on $G^*$, and (ii) $\mathring{w}(x) \leqq 0$ for all $\underline{x} \in G$.

Let $E = \{x;\ \mathring{w}(x) \nless 0,\ x \in \bar{G} \cap G^*\}$; $M$ is the largest invariant set in $E$.

**8.3.** LEMMA.

(i) $|W(y) - W(x)| \leqq \max_i |w_i(y) - w_i(x)|$.

(ii) *If* $w$ *is continuous at* $x^0 \in G^*$, *then so is* $W$.

(iii) *If* $w$ *is continuous on* $G^*$, *then given* $x \in G^*$ *there is a neighborhood* $N(x)$ *of* $x$ *such that* $y \in N(x)$ *implies* $J(y) \subset J(x)$.

(iv) *If* $w$ *is continuous on* $G^*$, *then, for* $y$ *sufficiently close to* $x$,

$$W(y) - W(x) = \max_{i \in J(x)} (w_i(y) - w_i(x)).$$

(v) *If* $w$ *is* $C^1$ *on* $G^*$, *then*

$$\dot{W}(x) = \max_{i \in J(x)} \dot{w}_i(x) = \max_i \mathring{w}_i(x).$$

*Proof.*

(i) Clearly, $W(y) - W(x) = \max_i [w_i(y) - W(x)] \leqslant \max_i [w_i(y) - w_i(x)] \leqq \max_i |w_i(y) - w_i(x)|$. Interchanging $x$ and $y$ establishes (i).

(ii) A direct consequence of (i).

(iii) An easy consequence of continuity.

(iv) For $y$ sufficiently close to $x$ we have $J(y) \subset J(x)$. But then

$$\begin{aligned} W(y) - W(x) &= \max_{i \in J(y)} [w_i(y)] - W(x) \\ &= \max_{i \in J(x)} [w_i(y) - W(x)] \\ &= \max_{i \in J(x)} [w_i(y) - w_i(x)]. \end{aligned}$$

(v) Let $\phi_i(t) = (1/t)(w_i(\pi(t, x) - w_i(x))$ with $\phi_i(0) = \dot{w}_i(x)$. Since each $w_i(x)$ is $C^1$, each $\phi_i(t)$ is continuous at $t = 0$. From (iv) we have that

$$\dot{W}(x) = \liminf_{t \to 0^+} \max_{i \in J(x)} \phi_i(t).$$

It then follows from (ii) that $\dot{W}(x) = \max_{i \in J(x)} \dot{w}_i(x)$. ■

From part (v) of the above lemma, we see that, if $w$ is a vector Liapunov function on $G$, then $W(x)$ is a scalar Liapunov function on $G$. Hence we have the following.

**8.4.** THEOREM. *If* (i) $w$ *is a vector Liapunov function of* (2.1) *on* $G$, *and* (ii) $\pi(t, x)$ *is precompact and in* $G$ *for all* $t \geqq 0$, *then* $\pi(t, x) \to M \cap W^{-1}(c)$ *as* $t \to \infty$ *for some* $c$.

In analogy to the discussion of the difference equation (10.2) in Chapter 1, we shall consider (2.1) in the form ($f(x)=A(x)x$)

$$\dot{x}=A(x)x, \tag{8.1}$$

where $A(x)$ is an $n\times n$ matrix function.

**8.5. Notation.** Let $B=(b_{ij})$ be an $n\times n$ matrix. Define $\hat{B}=(\hat{b}_{ij})$ by $\hat{b}_{ii}=b_{ii}$ and $\hat{b}_{ij}=|b_{ij}|$ for $i\neq j$.

We shall now establish and illustrate a result which refines and generalizes a condition for global asymptotic stability known to economists (see, for instance, [5]). Their condition is stated in terms of the Jacobian matrix $f'(x)$ of $f(x)$. The example we shall give motivates the manner in which the following result is stated (compare with Theorem 10.7 of Chapter 1).

**8.6.** THEOREM. *Let $G$ be an open set with $\bar{G}\subset G^*$ that is positively invariant relative to* (8.1). *Assume that there is a vector $c>0$ such that $\hat{A}(x)c<0$ for each $x\in G, x\neq 0$. Let $E_0$ be the set on the boundary of $G$ where $\hat{A}(x)c$ is not less than zero. Let $M_0$ be the largest invariant set in $E_0$. Then each solution of* (8.1) *starting in $\bar{G}$ approaches $\{0\}\cup M_0$ as $t\to\infty$.*

(i) *If $\{0\}\cup M_0=\{0\}$, then 0 is globally asymptotically stable relative to $\bar{G}$.*

(ii) *If $0\in G$, then each solution starting in $\bar{G}$ approaches either 0 or $M_0$. Hence, if no solution starting in $G$ approaches $M_0$, 0 is globally asymptotically stable relative to $G$.*

*Proof.* Let $w_i(x)=x_i^2/c_i^2$. For a fixed $x\in G$, $x\neq 0$, let $w_j(x)\geqq w_i(x)$, $i=1,\cdots,n$. Then a simple computation gives

$$\dot{w}_j(x)\leqq\frac{2|x_j|^2}{c_j}(\hat{A}(x)c)_j<0.$$

Therefore $w(x)$ is a vector Liapunov on $G$ (or $\bar{G}$) with $M=M_0\cup\{0\}$ if $0\in G$, or $M=M_0$ if $0\notin G$. Since $W(x)\to\infty$ as $\|x\|\to\infty$, each solution starting in $\bar{G}$ is bounded. Hence from Theorem 8.4 each solution starting in $G$ approaches $M_0\cup\{0\}$ as $t\to\infty$. In cases (i) and (ii) it follows that the origin is a global attractor relative to $G$. Stability relative to $G$ follows as in Theorem 7.8.

The next result shows how a vector Liapunov function can be used to establish instability.

**8.7.** THEOREM. *Let $N$ and $N_1$ be neighborhoods of the origin with $N\subset N_1$. If* (i) *$\hat{A}(x)=A(x)$ for all $x\in G=N\cap R_+^n$, and* (ii) *there is a $c>0$ such that $A(x)c>0$ for all $x\in G$, then the origin is unstable for* (8.1). *If, in addition,* (iii) *$G_1=N_1\cap R_+^n$ is positively invariant, then no solution starting in $G_1$ can approach the origin as $n\to\infty$, and the origin is strongly unstable.*

*Proof.* Let $w_i(x)=x_i/c_i$ and define $W(x)=\min_i w_i(x)$. If $w_j(x)\leqq w_i(x)$ for $x\in G$ and $i=1,\cdots,n$, then it follows easily from (i) and (ii) that $\dot{w}_j(x)\geqq(A(x)c)_j>0$. Now by (iv) of Lemma 8.3 (with "max" replaced everywhere by "min") we see that $\dot{W}(x)>0$ for all $x\in G$. The conclusions then follow from Theorem 7.12 with $V=-W$. ■

**8.9.** *Remark.* Note that Theorems 8.6 and 8.7 are unchanged if (8.1) is replaced by

$$\dot{x} = D(x)A(x)x, \tag{8.2}$$

where $D(x)$ is a diagonal matrix with positive diagonal for $x \in G$, $x \neq 0$.

**8.10.** *Exercise.* Let $A$ be a constant $n \times n$ matrix. $\hat{A} = A$ means that the off-diagonal terms are nonnegative. Economists call such matrices *Metzlerian* (for a complete discussion of matrices of this type, see [28]). Show that:

(a) $R_+^n$ is positively invariant (i.e., $e^{At} \geqq 0$ for all $t \geqq 0$) if and only if $\hat{A} = A$.

(b) If there is a vector $c > 0$ such that $\hat{A}c < 0$, then $A$ is stable ($e^{At} \to 0$ as $t \to \infty$).

(c) If $\hat{A} = A$ and there is a vector $c > 0$ such that $Ac \geqq 0$, then $A$ is not stable. If $Ac > 0$, then each solution of $\dot{x} = Ax$ starting in $R_+^n$ approaches infinity as $t \to \infty$.

(d) If $\hat{A} = A$, then the following are equivalent:

(i) $A$ is stable,

(ii) $A^{-1} \geqq 0$,

(iii) there is a vector $c > 0$ such that $Ac < 0$,

(iv) for each vector $c > 0$ there is an $i$ such that $(Ac)_i < 0$,

(v) there is a diagonal matrix $D$ with positive diagonal such that $DA + A^T D$ is negative definite.

The next example is the continuous version of Example 10.8 in Chapter 1.

**8.7.** *Example.* Consider (8.2) with

$$D(x) = \begin{pmatrix} \alpha_1 & 0 \\ 0 & \alpha_2 \end{pmatrix}, \quad A(x) = \begin{pmatrix} -\gamma_1 & 1 - x_1 \\ 1 - x_2 & -\gamma_2 \end{pmatrix};$$

i.e.,

$$\dot{x}_1 = \alpha_1(-\gamma_1 x_1 + (1 - x_1)x_2),$$
$$\dot{x}_2 = \alpha_2((1 - x_2)x_1 - \gamma_2 x_2).$$

Equations of this type and their generalizations to higher dimensions have been considered as epidemic models and predator-prey (or host-parasite) models. The result we shall establish is not new and for dimension $n = 2$ can be proved in many different ways. However, the method used here does not depend on the simple topology of the plane and can be generalized to higher dimensions. For the epidemic model, which is the one we consider, $\alpha_i$ and $\gamma_i$ are positive; $x_i$ is the fraction of the population $i$ infected and $(1 - x_i)$ is the fraction of the population susceptible (there is no immunity); $\alpha_i$ is a contact coefficient and $\alpha_i \gamma_i$ is a recovery coefficient. Here the $G$ of Theorem 8.6 is $G = \{x; 0 < x_i < 1, i = 1, 2\}$, and $G$, and hence $\bar{G}$, is positively invariant. Note also that $\hat{A}(x) = A(x)$ for $x \in G$. Since

$$A(x)\begin{pmatrix} 1 + \gamma_2 \\ 1 + \gamma_1 \end{pmatrix} = \begin{pmatrix} 1 - \gamma_1\gamma_2 - (1 + \gamma_1)x_1 \\ 1 - \gamma_1\gamma_2 - (1 + \gamma_2)x_2 \end{pmatrix},$$

the conditions of Theorem 8.6 are satisfied when $\gamma_1\gamma_2 \geqq 1$. At worst (when $\gamma_1\gamma_2 = 1$) $E_0$ is the intersection of the coordinate axes with $\bar{G}$, but $M_0 = \{0\}$. Therefore, if $\gamma_1\gamma_2 \geqq 1$, the origin is globally asymptotically stable relative to $\bar{G}$. Recovery is sufficiently rapid compared to contact that the infection tends to die out; in fact, since (see the proof of Theorem 8.6) $w_1(x) = x_1^2/(1+\gamma_2)^2$ and $w_2(x) = x_2^2/(1+\gamma_1)^2$, $\max(x_1/(1+\gamma_2), x_2/(1+\gamma_1))$ is always decreasing.

If $\gamma_1\gamma_2 < 1$, then near the origin in $G$ we have

$$A(x)\begin{pmatrix} 1+\gamma_2 \\ 1+\gamma_1 \end{pmatrix} > 0.$$

Hence, by Theorem 8.7 the origin is unstable and no solution starting in $G$ can approach the origin as $n \to \infty$. (In fact, it is not difficult to show that each solution starting at a point of $\bar{G}$ other than the origin has its positive limit set in $G$. We see later that this is true.) There is an equilibrium point in $G$ at $x_1^0 = (1-\gamma_1\gamma_2)/(1+\gamma_1)$ and $x_2^0 = (1-\gamma_1\gamma_2)/(1+\gamma_2)$. Letting $u_1 = x_1 - x_1^0$ and $u_2 = x_2 - x_2^0$, we have in these coordinates

$$\dot{u} = DB(x)u, \tag{8.3}$$

where

$$B(x) = \begin{pmatrix} -\rho & 1-x_1 \\ 1-x_2 & -\rho^{-1} \end{pmatrix}$$

and $\rho = (1+\gamma_1)/(1+\gamma_2)$. Since $\hat{B}(x) = B(x)$ for $x$ in $G$ and

$$B(x)\begin{pmatrix} 1+\gamma_2 \\ 1+\gamma_1 \end{pmatrix} = -\begin{pmatrix} (1+\gamma_1)x_1 \\ (1+\gamma_2)x_2 \end{pmatrix},$$

the conditions of Theorem 8.6 are satisfied for (8.3). It is easily seen that $M_0$ is $u = -x^0$ $(x = 0)$. Since no solution starting in $G$ can approach $u = -x^0$, the equilibrium point $u = 0$ $(x = x^0)$ is globally asymptotically stable relative to $G$. In fact, every solution starting in $\bar{G}$ other than at $x = 0$ enters $G$ and therefore approaches $x = x^0$ as $t \to \infty$. The system tends to a stable level of infection as $t \to \infty$ when $\gamma_1\gamma_2 < 1$. Here $\max\{|x_1 - x_1^0|/(1+\gamma_2), |x_2 - x_2^0|/(1+\gamma_1)\}$ is always decreasing. In this case, unlike the discrete model (Example 10.8 of Chapter 1), we obtain a complete answer. In the discrete model we were not able to establish the global asymptotic stability of $x^0$ for all $\gamma_1\gamma_2 < 1$.

# CHAPTER 3

## Functional Differential Equations. Local Semidynamical Systems

**1. Introduction.** Difference-differential equations have been studied almost as long as ordinary differential equations but it was only around fifteen years ago that the geometric approach, so successful for ordinary differential equations, was adopted. This point of view was taken by Krasovskii in [50] in his development of Liapunov's Second or Direct Method for equations with delays. These equations define flows (local semidynamical systems) in a space of functions, and recognizing this brought a great impetus to the development of the theory of difference-differential equations by analogy to the more highly developed theory of ordinary differential equations. This enabled Krasovskii to extend Liapunov's classical theory to functional differential equations and led Hale [36] in 1963 (see also [37], [40]) to apply the invariance principal to such systems. This generalization of Liapunov's Direct Method has implications for functional differential equations and beyond, that exceed those of the corresponding developments for ordinary differential equations. This we shall now present in the broader context of local semidynamical systems. While time does not permit us to go beyond autonomous retarded functional differential equations, the applications today are far more extensive (see the references given at the end of this chapter).

Examples of retarded functional differential equations are difference-differential equations

$$\dot{x}(t) = f(x(t), x(t-1)) \tag{1.1}$$

and integral-differential equations

$$\dot{x}(t) = \int_{t-1}^{t} f(x(s), s)\, dx.$$

In each of the above the state of the system at time $t$ depends upon a portion of its past history (hereditary systems). Such equations arise in many different applications. A good introduction to the general theory is [40].

**2. Autonomous retarded functional differential equations**

**2.1. Notation.**

$x: [-r, a) \to R^n$, $r$ and $a$ positive.

$C = C([-r, 0], R^n)$ is the space of continuous functions $\phi: [-r, 0] \to R^n$. $C$ is a Banach space with $\|\phi\| = \max_{t\in[-r,0]} \|\phi(t)\|$, and convergence in $C$ is uniform convergence on $[-r, 0]$.

The function $x_t:[-r,0]\to R^n$ is defined by $x_t(0)=x(t+0)$, $-r\leqq 0\leqq 0$ and $0\leqq t<a$; $x_t$ is a portion at $t$ of the past history of $x(t)$.

$f:X\subset C\to R^n$, where $X$ is an open set in $C$ and $f$ is continuous. Unless we say otherwise, topological concepts are relative to $C$.

An autonomous retarded differential equation is then

$$\dot{x}(t)=f(x_t). \tag{2.1}$$

**2.2.** Definition. A function $x:[-r,a)\to R^n$ is a *solution of* (2.1) if, for some $a>0$, $x$ satisfies (2.1) for all $t\in[0,a)$; $x(t)$ is *a solution of the initial value problem*

$$\dot{x}(t)=f(x_t), \qquad x_0=\phi \tag{2.2}$$

if $x(t)$ is a solution of (2.1) on $[-r,a)$ for some $a>0$ and $x(t)=\phi(t)$ for $-r\leqq t\leqq 0$, $\phi\in X$. We limit ourselves to continuous initial data.

We shall always assume that each initial value problem (2.2) has a unique solution $x(t)$ defined for $t\in[-r,\omega(\phi))$, $\omega(\phi)>0$, where $[-r,\omega(\phi))$ is the maximum interval of definition. We also assume throughout that $f$ maps bounded sets of $X$ into bounded sets of $C$. These assumptions are satisfied if $f$ is assumed to be locally Lipschitz continuous on $X$. The existence-uniqueness theorems and the theorems on continuous dependence are similar to those for ordinary differential equations (see [40]), except—as for difference equations—existence and uniqueness are only in the forward direction of time.

**3. The flow defined by (2.1).** Let $x(t)$ be the unique solution of (2.2), let $I(\phi)=[0,\omega(\phi))$, and define $\chi=\{(t,\phi);\ t\in I(\phi),\ \phi\in X\}\subset R_+\times X$. The mapping $\pi:\chi\to X$ is defined by $\pi(t,\phi)=x_t$; $\pi(t,\phi)$ is the *motion* or *flow* on $X$ defined by the solutions of (2.1). Note that $x(t)=x_t(0)=\pi(t,\phi)(0)$.

**3.1.** Definition. A function $\psi\in C$ is a *positive limit point* of $\pi(r,\phi)$ if there is a sequence $t_n\in I(\phi)$ such that $t_n\to\omega(\phi)$ and $\pi(t_n,\phi)\to\psi$ as $n\to\infty$. The *positive limit sét* $\Omega(\phi)$ is the set of all positive limit points of $\pi(t,\phi)$. Note that $\pi(t,\phi)$ can have positive limit points on the boundary of $X$.

The basic properties of solutions of (2.1) (see [40]) can then be summarized as follows in terms of $\pi$:

$P_1$. Either $I(\phi)=[0,\infty)$ or $\Omega(\phi)\cap X$ is empty.

$P_2$. $\pi(0,\phi)=\phi$.

$P_3$. $s,t\in R_+=[0,\infty)$ and $s+t\in I(\phi)$ imply $s\in I(\pi(t,\phi))$ and $\pi(s,\pi(t,\phi))=\pi(s+t,\phi)$.

*$P_4$*. $I(\phi)$ is lower semicontinuous and $\pi$ is continuous; i.e., if $\phi\in X$ and $t\in I(\phi)$, then $(t_n,\phi_n)\to(t,\phi)$ implies $t\in I(\phi_n)$ for all $n$ sufficiently large and $\pi(t_n,\phi_n)\to\pi(t,\phi)$ as $n\to\infty$.

A mapping $\pi$ with the above properties is called a *local semidynamical system* (see [13]). It is "local" because of the local existence of solutions and "semi" because solutions are defined only for positive $t$. $P_1$ is the maximality property of $I(\phi)$, $P_3$ is the semigroup property and is equivalent to uniqueness in the forward direction of time (Exercise 3.2), and $P_4$ is continuous dependence on initial values.

**3.2.** *Exercise.* Show that: (a) $\omega(\phi)=t+\omega(\pi(t,\phi))$ for all $t\in I(\phi)$, and (b) $t\in I(\phi)\cap I(\psi)$ and $\pi(t,\phi)=\pi(t,\psi)$ implies $I(\phi)=I(\psi)$ and $\pi(t+s,\phi)=\pi(t+s,\psi)$ for all $s\geqq 0$ and $t+s\in I(\phi)$.

**4. Invariance.**

**4.1.** DEFINITION. A set $H\subset X$ is said to be *invariant* with respect to (2.1) if $I(\phi)=[0,\infty)$ for each $\phi\in H$ and $\pi(t,H)=H$ for all $t\geqq 0$.

**4.2.** *Exercise.* Let $I(\phi)=[0,\infty)$. A function $\pi^*(\cdot,\phi)$ defined on $(-\infty,\infty)$ to $X$ is called an *extension* of $\pi(t,\phi)$ if (i) $\pi^*(0,\phi)=\phi$ and (ii) $\pi(s,\pi^*(t,\phi))=\pi^*(t+s,\phi)$ for all $s\in[0,\infty)$ and all $t\in(-\infty,\infty)$; $x^*(t)=\pi^*(t,\phi)(0)$ is called an *extension of the solution* $x(t)=\pi(t,\phi)(0)$. Show, using the Axiom of Choice, that: $H$ is invariant if and only if, for each $\phi\in H$, $I(\phi)=[0,\infty)$ and $\pi(t,\phi)$ has an extension $\pi^*(t,\phi)$ contained in $H$ for all $t$.

**4.3.** DEFINITION. Let $x(t)$ be a solution of (2.1) satisfying $x_0=\phi$, and define $\gamma^+(\phi)=\pi(I(\phi),\phi)=\{x_t;\, t\in I(\phi)\}$, called the *positive trajectory from* $\phi$. The solution $x(t)$ is said to be *precompact* if $\overline{\gamma^+(\phi)}$ is compact and $\overline{\gamma^+(\phi)}\subset X$ (i.e., $\gamma^+(\phi)$ is precompact relative to $X$, and this is equivalent to $\gamma^+(\phi)$ is precompact relative to $C$ and $\Omega(\phi)\subset X$). Note that, if $x(t)$ is precompact, then $I(\phi)=[0,\infty)$.

Now, just as before, it is not difficult to prove that the following theorem holds.

**4.4.** THEOREM. *If $x(t)$ is a precompact solution of* (2.1), *then its positive limit set $\Omega(\phi)$ is contained in $X$, and is nonempty, compact, invariant, connected and $x_t\to\Omega(\phi)$ as $t\to\infty$.*

**4.5. Notation.** $V: X\to R$. Relative to (2.1),

$$\dot V(\phi)=\liminf_{t\to 0^+}\frac{1}{t}[V(\pi(t,\phi))-V(\phi)].$$

**4.6.** DEFINITION. Let $G$ be a subset of $X$. $V$ is said to be a *Liapunov function of* (2.1) *on* $G$ if (i) $V$ is continuous on $X$ (or $\bar G$), and (ii) $\dot V(\phi)\leqq 0$ for all $\phi\in G$.

If $V$ is a Liapunov function, then $E=\{\phi;\, \dot V(\phi)=0,\, \phi\in\bar G\}$ and $M$ is the largest invariant set in $E$. We then have, as before, the same extension of Liapunov's direct method and all of the corresponding theorems on stability and instability. We state only the principal result.

**4.7.** THEOREM. *Let $V$ be a Liapunov function of* (2.1) *on $G$, and let $x(t)$ be a solution of* (2.1) *that is precompact and such that $x_t$ remains in $G$ for all $t\geqq 0$. Then, for some $c$, $x_t\to M\cap V^{-1}(c)$.*

This is fine. Everything is as before but for applications there remains a difficulty, which arises with each extension of the invariance principle when the state space $X$ is not locally compact (see, for instance, [38]). As a practical matter, the question is: How, given the differential equation (2.1), do you verify the precompactness of solutions? Boundedness of solutions can be determined directly from the differential equations (2.1) by Liapunov functions and for retarded functional differential equations the answer is given in the following lemma (remember we are assuming that $f$ maps bounded sets into bounded sets).

**4.8.** LEMMA. *Let $x(t)$ be a bounded solution of* (2.1). *If $x_t$ has no positive limit points on the boundary of $X$, then $x(t)$ is precompact.*

*Proof.* Let $K$ be an upper bound of $\|f(x_t)\|$ for $t\in I(\phi)$, and $\gamma^+(\phi)=\{x_t;\, t\in I(\phi)\}$. If $t,\, t+\theta\in I(\phi)=[0,\omega(\phi))$, then

$$\|x(t+\theta)-x(t)\|=\left\|\int_t^{t+\theta} f(x_s)\,ds\right\|\leqq K|\theta|$$

and $x(t)$ is uniformly continuous on $[0,\omega(\phi))$. Since, for any $0<a<\omega(\phi)$, $x(t)$ is uniformly continuous on $[-r,a]$, $x(t)$ is uniformly continuous on $[-r.\,\omega(\phi))$. Therefore the family $\gamma^+(\phi)$ of functions is equicontinuous on $[-r,0]$, and $\gamma^+(\phi)$ is precompact in $C$; since $\gamma^+(\phi)\subset X$, $x(t)$ is precompact. ■

We shall now look at one example, which, while simple, illustrates how this extension of Liapunov's direct method can be applied. This same equation was studied in [40], but note that we obtain here more information on the asymptotic behavior of solutions.

**4.9.** *Example.* Consider the difference-differential equation

$$\dot{x}(t)=ax^3(t)+bx^3(t-r),\qquad r>0.$$

Note that, when $r=0$, we have global asymptotic stability if $a+b<0$ and every solution except the trivial one is unbounded when $a+b>0$.

Let

$$V(\phi)=-\frac{1}{2a}\phi^4(0)+\int_{-r}^0 \phi^6(\theta)\,d\theta.$$

Since $V(x_t)=-(1/2a)x^4(t)+\int_{t-r}^t x^6(s)\,ds$,

$$\dot{V}(\phi)=-\left(\phi^6(0)+\frac{2b}{a}\phi^3(0)\phi^3(-r)+\phi^6(-r)\right).$$

Hence, if $|b|\leqq|a|$, $V$ is a Liapunov function on $C$. If $a<0$, $V$ is positive definite and $V\to\infty$ as $\|\phi\|\to\infty$, and every solution is bounded and hence precompact.

*Case* 1. $a<0$ and $|b|<|a|$. $E$ is the set of functions $\phi$ continuous on $[-r,0]$ satisfying $\phi(0)=\phi(-r)=0$, but $M=\{0\}$, the zero function on $[-r,0]$ (the origin in $C$). Hence the origin is globally asymptotically stable.

*Case* 2. $a<0$ and $b=a$. The set $E$ consists of functions satisfying $\phi(0)=-\phi(-r)$. If the flow remains in $E$, then $\dot{x}(t)=0$. Hence, $M$ corresponds to constant functions $\phi=c$, but then $c=0$ and $M=\{0\}$. Again the origin is globally asymptotically stable.

*Case* 3. $a<0$ and $b=-a$. Here $E$ corresponds to $\phi(0)=\phi(-r)$ and $M$ corresponds to $\phi=c_0$ (each constant function is an equilibrium point). The intersection of $M$ and $V^{-1}(c)$ is a finite number of constant functions. Since $\Omega$ is connected, each motion $x_t$ approaches a constant function.

*Case* 4. $a>0$ and $|b|<a$ (or $b=a$). The set $G=\{\phi;\ V(\phi)<0\}$ is nonempty and positively invariant, and $M$ is the origin. No motion $x_t$ starting in $G$ can approach $M$ which is on the boundary of $G$, and therefore every solution starting in $G$ is unbounded; in fact, one can show for each solution $x(t)$ starting in $G$ that $|x_t|\to\infty$ in finite time or as $t\to\infty$.

More refined results depending on $r$ would require the use of a more sensitive Liapunov function. For other examples, see [40].

Today the invariance principle has been extended to neutral functional differential equations, integral equations, and to certain classes of partial differential equations and evolutionary equations. This is beyond the scope of these lectures. (For a discussion of such extensions and their applications, see [16]–[18], [22], [24], [25], [38], [41], [44], [59], [76]–[80], [89], [97], [98], [101]–[104], [107]–[109].

# CHAPTER 4

## Abstract Discrete Dynamical Systems and Processes. Nonautonomous Difference Equations

**1. Introduction.** This chapter has two objectives. The first is to develop in outline invariance principles for nonautonomous difference equations. Here we follow Miller, Sell, Dafermos, and Artstein (see Appendix A for an historical account of these developments) and borrow heavily both their ideas and their terminology. At this point one might go directly to Appendix A where the discussion is less formal and more detailed. There are, however, some differences between ordinary differential equations and difference equations that show up in this chapter and here the theory is really more elementary. The point of view here is also slightly different from Appendix A.

The second objective is to demonstrate that basically the theory is quite simple and that the only structure required on the state space is a concept of convergence. This structure need not even be topological. We carry this development only far enough to be able to discuss what is involved in obtaining invariance properties of the solutions (motions) of nonautonomous systems (processes). For abstract continuous dynamical systems with a topological structure on the state space, see [13]. Recent developments, such as those of Sell in [97] and Miller and Sell [77]–[80], place topological dynamics in a new light. This chapter and Appendix A give a reason for considering "nontopological" dynamics. For the most part the proofs follow so closely those of Chapter 1 that it suffices simply to give an outline of the theory.

**2. Discrete dynamical systems. Autonomous difference equations**

**2.1.** DEFINITION. A set $P$ is a *Fréchet* (or *convergence*) *space* if each sequence $p_n$ in $P$ either has a limit $p$ in $P$ associated with it or not; that is, each sequence either converges or diverges. If $p_n$ has a limit, we denote this by $p_n \to p$ as $n \to \infty$ (or, simply, by $p_n \to p$). This convergence has the following properties:

(i) $p_n \to p$ and $p_n \to q$ implies $p = q$;

(ii) $p_n = p$, $n = 1, 2, \cdots$, implies $p_n \to p$;

(iii) $p_n \to p$ implies $p_{n_i} \to p$ for each subsequence $p_{n_i}$ of $p_n$.

These are the spaces of class $(L)$ studied by Fréchet in his dissertation around 1906 (see [51, Chap. 21]).

Continuity, compactness, closure, etc. are defined in the usual way with respect to convergence. They are all sequential concepts. For instance, $K \subset P$ is *compact* if each sequence in $K$ has a convergent subsequence whose limit is in $K$. A compact set is closed but the closure of a set may not be closed. A set $H$ is *precompact* if it is contained in a compact set. A precompact set is compact if and only if it is closed.

**2.2.** DEFINITION. Let $P$ be a Fréchet space. A map $\pi: J_+ \times P \to P$ is a *discrete dynamical system* if:

$D_1$. $\pi(0, p) = p$, all $p \in P$.

$D_2$. $\pi(n, \pi(k, p)) = \pi(n+k, p)$, all $n, k \in J_+$ and $p \in P$.

$D_3$. $\pi$ is continuous.

Continuity of $\pi$ means $p_k \to p$ implies $\pi(n, p_k) \to \pi(n, p)$ for each $n \in J_+$ ($J_+ \times P$ is a Fréchet space). A mapping $\pi$ satisfying $D_1$ and $D_2$ will be called a *flow* (or *semi-flow*, if there is a reason for the distinction).

With each discrete dynamical system there is an associated difference equation

$$p' = T(p), \tag{2.1}$$

where $T(p) = \pi(1, p)$. Conversely, each continuous $T: P \to P$ has the dynamical system $\pi$, $\pi(n, p) = T^n p$ associated with it. Limit points, limit sets, invariance, etc., are as defined in Chapter 1. Boundedness of $T^n x$ is replaced by $T^n p$ precompact; i.e., $\pi(J_+, p)$ is precompact. This is often called *positive stability in the sense of Lagrange*.

We now follow the developments of Chapter 1 with, of course, some weakening of the results because of the general nature of the state space $P$.

**2.3.** PROPOSITION. Every limit set $\Omega(p)$ is positively invariant. If $T^n p$ is precompact, then $\Omega(p)$ is nonempty, precompact, and invariant.

**2.4.** *Exercise.* Give an example of a discrete dynamical system where $T^n p$ is precompact and $\Omega(p)$ is not closed (i.e., not compact).

**3. An invariance principle.** The concept of a Liapunov function $V: P \to R$, $\dot{V}$, and the associated sets $E$ and $M$ are as before with $R^m$ replaced by $P$. In place of $T^n p = \pi(n, p) \to S$ we have $\Omega(p) \subset S$.

**3.1.** THEOREM (invariance principle). *Let $\pi$ be a discrete dynamical system on $P$, and let $V$ be a Liapunov function for $\pi$ on $G$. If $\pi(J_+, p)$ is precompact and $\pi(J_+, p) \subset G$, then $\Omega(p) \subset M \cap V^{-1}(c)$ for some $c = c(p)$.*

With this invariance principle we then have, as before, a direct method for studying stability and instability. Some modifications are required, since here we assume no topology on $P$, and everything must be defined in terms of convergence. For instance, Chapter 1, Lemma 7.4 tells us how to define stability. We shall not explore this here. There are many interesting results but without some applications in mind it is difficult to determine a priori their significance. Rather let us proceed and see why these abstract dynamical systems are of interest.

**4. Nonautonomous difference equations. Discrete processes.**

**4.1. Notation.** We consider the nonautonomous difference equation

$$x' = T(n, x), \tag{4.1}$$

where $T: J \times X \to X$, $X$ is a Fréchet space and $T$ is continuous.

$T_{(n)}: X \to X$ is defined by $T_{(n)}x = T(n, x)$.

$T_k: J \times X \to X$, a *translate* of $T$, is defined by

$$T_k(n, x) = T(n+k, x).$$

For each $n_0 \in J$ and $n \in J_+$, $T^n_{n_0}: X \to X$ is defined by $T^0_{n_0} = I$ and $T^{n+1}_{n_0} = T_{(n+n_0)} T^n_{n_0}$. $\hat{T}: J_+ \times J \times X \to X$ is defined by $\hat{T}(n, n_0, x) = T^n_{n_0} x$. $T^n_{n_0} = \prod_{j=0}^{n-1} T_{(n_0+j)}$, the composition of $n$ functions.

Note that:

$$T^n_{n_0+k} T^k_{n_0} = T^{n+k}_{n_0} \tag{4.2}$$

and

$$\hat{T}_k(n, n_0, x) = \hat{T}(n, n_0+k, x). \tag{4.3}$$

Equation (4.2) is the uniqueness of the solution in the positive direction of time. Equation (4.3) expresses the fact that, if $x$ is a solution of (4.1), then $x(n+k)$ is a solution of $x' = T(n+k, x) = T_k(n, x)$.

**4.2.** Definition. A map $\hat{T}: J_+ \times J \times X \to X$ is called a *discrete process* on $X$ if:

$P_1$. $\hat{T}(0, n_0, x) = x$ for each $n_0 \in J$ and $x \in X$.

$P_2$. $\hat{T}(n, n_0+k, \hat{T}(k, n_0, x)) = \hat{T}(n+k, n_0, x)$ for all $n, k \in J_+$, $n_0 \in J$ and $x \in X$.

$P_3$. $\hat{T}$ is continuous.

With each nonautonomous difference equation (4.1) the $\hat{T}$ defined above is a discrete process. Conversely, if $\hat{T}$ is a process the difference equation associated with it is (4.1) with $T(n, x) = \hat{T}(1, n, x)$. If $\phi(n, n_0, x^0, T)$ is the solution of (4.1) satisfying $\phi(n_0, n_0, x^0, T) = x^0$, then $\phi(n+n_0, n_0, x^0, T) = \hat{T}(n, n_0, x^0) = T^n_{n_0} x^0$. If the state of the system at time $n_0$ is $x^0$, then its state at time $n+n_0$ is $T^n_{n_0} x^0$.

**5. Dynamical systems associated with nonautonomous difference equations. Skew-product flows.** Let $W$ denote the set of all functions $T: J \times X \to X$. Let $P = X \times W$. Given $p = (x, T) \in P$ we define for each $n \in J_+$ and a fixed $n_0 \in J$,

$$\pi(n, p) = (T^n_{n_0} x, T_n) = (\hat{T}(n, n_0, x), T_n). \tag{5.1}$$

It is easy to see that $\pi$ is a flow on $P$. This is the *skew-product flow*.

Suppose that a convergence has been defined on $W$ and that $W_0 \subset W$ is a Fréchet space (see Exercise 5.7). Then $P_0 = X \times W_0$ is a Fréchet space with convergence defined in the usual way for a product space. For a stability theory we would like $\pi$ to be a discrete dynamical system and to have the invariance property of the limit sets in $P_0$ of precompact motions $\pi(n, p)$ $(p = (x, T))$ induce an invariance for the limit sets in $X$ of the precompact motions $T^n_{n_0} x$.

Let $H(T)=\overline{T_{J+}}$, called the (positive) *hull* of $T$. It is the closure in $W$ of the positive translates of $T$. We want therefore to have $H(T)\subset W_0$ and to have $\pi$ be a discrete dynamical system on $X\times H(T)$. Mappings $T$ in $W_0$ with this property are said to be *regular*. This concept of regularity is relative to the convergence defined on $W$.

Let $\phi(n)=T_{n_0}^{n-n_0}x^0$ be any solution of $x'=T(n,x)$. Suppose there is a sequence $n_i$ such that $\phi(n_i)\to y$ and $T_{n_i}\to S$. Then $y\in\Gamma$, the limit set of $\phi(n)$, and $S\in H_\infty(T)$, the limit set of the translates $T_n$; $H_\infty(T)$ is called the *asymptotic hull of* $T$. The equation $x'=S(n,x)$ is called a *limiting equation* of $x'=T(n,x)$. If $T$ is regular, then the limit set $\Omega$ of $\pi(n,p)$, $p=(x^0,T)$, is positively invariant (Proposition 2.3). Now $q=(y,S)\in\Omega$ implies $\pi(n,q)\in\Omega$ for all $n\geqq 0$; i.e., $S_{n_0}^{n}x\in\Gamma$ for all $n\geqq 0$. The functions $T$ in $W_0$ with the property that $T_n$ is precompact (i.e., $T_{J_+}$ is precompact in $W_0$), are said to be *compact*. The positive invariance of $\Omega$ then gives us the following induced invariance property for $\Gamma$ (in Appendix A this is called *semi-quasi-invariance*).

**5.1.** THEOREM (induced invariance). *Assume that $T$ is regular and compact, and let $\Gamma$ be the limit set of a solution $\phi(n)$ of $x'=T(n,x)$ defined for all $n\geqq n_0$. If $y\in\Gamma$, then there is an $S\in H_\infty(T)$ and a solution $\psi(n)$ of the limiting equation $x'=S(n,x)$ satisfying $\psi(n_0)=y$ and $\psi(n)\in\Gamma$ for all $n\geqq n_0$.*

**5.2.** DEFINITION. Given a $T\in W$, an $x\in X$, and an $n_0\in J$, $T_{n_0}^{n}x$ is said to have *an extension $T_{n_0}^{(n)}x$ on $J$* if $T_{n_0}^{(0)}x=x$ and $T_{n_0+n}^{1}T_{n_0}^{(n)}x=T_{n_0}^{(n+1)}x$. Note, of course, this is an extension—$T_{n_0}^{(n)}x=T_{n_0}^{n}x$ for $n\in J_+$. If $\phi(n)=T_{n_0}^{n-n_0}x$ is any solution of (4.1) defined for $n\geqq n_0$, then $\phi^*(n)=T_{n_0}^{(n-n_0)}x$ is an *extended solution of* (4.1) *on* $J$—$\phi^*(n)=\phi(n)$ for all $n\geqq n_0$.

Under the additional assumption that $\phi(n)$ is precompact we have the following two results, since then the limit set $\Omega$ is invariant (Proposition 2.3).

**5.3.** THEOREM (a stronger induced invariance). *If, in addition to the assumptions of Theorem* 5.1 *the solution $\phi(n)$ of $x'=T(n,x)$ is precompact, then there is, given $y\in\Gamma$, an $S\in H_\infty(T)$ and an extended solution $\psi(n)$ of the limiting equation $x'=S(n,x)$ satisfying $\psi(n_0)=y$ and $\psi(n)\in\Gamma$ for all $n\in J$.*

**5.4.** *Remark.* Since $H_\infty(T)$ is invariant under translations, nothing is lost by taking $n_0=0$.

**5.5.** THEOREM (another induced invariance). *Under the assumptions of Theorem* 5.3, *there is, given $S\in H_\infty(T)$, a $y\in\Gamma$ and an extended solution $\psi(n)$ of $x'=S(n,x)$ satisfying $\psi(n_0)=y$ and $\psi(n)\in\Gamma$ for all $n\in J$.*

**5.6.** *Remark.* An important special case of Theorem 5.5 is when $\phi(n)\to y$ as $n\to\infty$. Then $y$ is an equilibrium point of each limiting equation $x'=S(n,x)$; i.e., $S(n,y)=y$ for each $n\geqq n_0$.

**5.7.** *Exercise.* Define convergence in $W$ by $S^k \to S$ if $x^k \to x$ implies $S^k(n, x^k) \to S(n, x)$ for each $n \in J$. Then show that the maximal $W_0$ is the set of continuous functions in $W$ and that each $T$ in $W_0$ is regular.

**5.8.** *Exercise.* Interpret the above results for a periodic difference equation $x' = T(n, x)$ ($T(n+k, x) = T(n, x)$ for all $n \in J$; i.e., $T_k = T$).

**6. Finite-dimensional nonautonomous difference equations.** Here we consider the difference equation

$$x' = T(n, x), \tag{6.1}$$

where

$$T: J \times R^m \to R^m.$$

Thus $X = R^m$ and $W$ is the set of all functions mapping $J \times R^m$ into $R^m$. What we want to do is to introduce a convergence on $W$ that results in a large class of functions $T$ that are both compact and regular.

Assume:

$H_1$. $T(n, x)$ is bounded on $J_+$ for each $x \in R^m$ (pointwise boundedness).

$H_2$. $T(n, x)$ is equicontinuous ($n \geqq 0$) on compact sets of $X$.

Consider the following three types of convergence on $W$, the set of all functions $T: J \times R^m \to R^m$:

$C_1$. $S^k \to S$ if $S^k(n, x) \to S(n, x)$ for each $(n, x) \in J_+ \times R^m$.

$C_2$. $S^k \to S$ if $x^k \to x$ implies $S^k(n, x^k) \to S(n, x)$ for each $n \in J_+$.

$C_3$. $S^k \to S$ if the convergence is uniform on compact sets of $J_+ \times R^m$.

**6.1.** *Exercise.* Show that: If $T$ satisfies $H_1$ and $H_2$, then with respect to $T_{n_i} \to S$ (the convergence of sequences of translates of $T$) the three convergences $C_1$, $C_2$ and $C_3$ are equivalent.

**6.2.** LEMMA. *If $T$ satisfies* $H_1$ *and* $H_2$, *then $T$ is regular and compact* (*relative to any of the convergences* $C_1$, $C_2$ *or* $C_3$).

**7. Liapunov functions.** For $V: J \times R^m \to R$ we define, relative to (6.1),

$$\dot{V}(n, x) = V(n+1, T(n, x)) - V(n, x)$$

so that, if $x(n)$ is a solution of (6.1),

$$\dot{V}(n, x(n)) = V(n+1, x(n+1)) - V(n, x(n)).$$

**7.1.** DEFINITION. $V: J \times R^m \to R$ is called a *Liapunov function* for (5.1) *on $G$* if (i) $V$ is continuous, (ii) given $y \in R^m$ there is a neighborhood $N$ of $y$ and an $\alpha > -\infty$ such that $V(n, x) \geqq \alpha$ for all $x \in N$ and all $n$ sufficiently large, (iii) there is a continuous function $W: R^m \to R$ such that $\dot{V}(n, x) \leqq -W(x) \leqq 0$ for all $x \in G$ and all $n$ sufficiently large.

The set $E$ is defined by

(7.1) $$E=\{x;\ W(x)=0,\ x\in\bar{G}\}.$$

**7.2.** THEOREM. *Let $V$ be a Liapunov function of* (6.1) *on $G$. If a solution $\phi(n)$ of* (6.1) *is bounded and remains in $G$ for all $n\geqq n_0$, then $\phi(n)\to E$ as $n\to\infty$.*

**7.3.** *Remark.* It is assumed here that $T$ is continuous. If $V(n,x)=V(x)$, the conclusion of Theorem 7.2 becomes: *for some $c$, $\phi(n)\to E\cap V^{-1}(c)$.*

**7.4.** *Example.* Consider $x''+a(n)x=0$. The equivalent system is $x'=y$, $y'=-a(n)x$. Let $V(x,y)=x^2+y^2$. Then $\dot{V}(x,y)=-(1-a^2(n))x^2$. Hence, if $|a^2(n)|\leqq\alpha<1$ for all $n\geqq n_0$, each solution $(x(n),y(n))$ approaches a point $(0,b)$ as $n\to\infty$. The next theorem will enable us to conclude that the origin is globally asymptotically stable.

Relative to a Liapunov function $V$ we define $M$ to be the largest set in $E$ having the strong induced invariance property of Theorem 5.3; i.e., $x\in M$ if $x\in E$ and if for some $S\in H_\infty(T)$ there is an extended solution of $x'=S(n,x)$ starting at $x$ that remains in $E$ for all $n\in J$. We then have from Theorem 5.3 the following.

**7.5.** THEOREM. *Let $V$ be a Liapunov function of* (6.1) *on $G$, and assume that $T$ satisfies* $H_1$ *and* $H_2$. *If a solution $\phi(n)$ of* (6.1) *is bounded and in $G$ for $n\geqq n_0$, then $\phi(n)\to M$ as $n\to\infty$. If $V(n,x)=V(x)$, then, for some $c$, $\phi(n)\to M\cap V^{-1}(c)$.*

**7.6.** *Example.* Returning to Example 7.4 we see from Remark 5.6 that if $(x(n),y(n))\to(0,b)$ as $n\to\infty$, then $(0,b)$ must be an equilibrium point of each limiting equation (convergence is $C_1$, $C_2$ or $C_3$). But under our assumptions the only such point is the origin ($M=\{0\}$) and the origin is globally asymptotically stable.

**7.7.** *Exercise.* Let $A(n)$ be a bounded $m\times m$ real matrix-valued function on $J_+$, and consider the difference equation $x'=A(n)x$. Assume that there is a positive definite matrix $Q$ and a positive semidefinite matrix $B$ such that (a) $Q-A^T(n)QA(n)-B$ is positive semidefinite for each $n\geqq 0$, and (b) if $n_i\to\infty$ and $A(n+n_i)\to A_\infty(n)$ for each $n\geqq 0$, then $Bx=0$ and $BA_\infty(0)x=0$ implies $x=0$. Conclude that the origin is globally asymptotically stable.

# References


[1] M. A. AIZERMAN, *On a problem concerning global stability of dynamical systems*, Uspekhi Mat. Nauk, 4 (1949), pp. 187–188.

[2] M. A. AIZERMAN AND F. R. GANTMACHER, *Absolute Stability of Regulator Systems*, Holden-Day, San Francisco, Calif., 1964.

[3] B. ANDERSON AND J. MOORE, *Linear Optimal Control*, Prentice-Hall, Englewood Cliffs, N.J., 1971.

[4] K. J. ARROW, H. D. BLOCK AND L. HURWICZ, *On the stability of competitive equilibrium. II*, Econometrica, 27 (1959), pp. 82–109.

[5] K. J. ARROW AND F. H. HAHN, *General Competitive Analysis*, Holden-Day, San Francisco, Calif. 1971.

[6] Z. ARTSTEIN, *Continuous dependence of fixed points of condensing maps*, Dynamical Systems. An International Symposium, Academic Press, New York, 1976.

[7] ———, *Continuous dependence of solutions of Volterra integral equations*, SIAM J. Math. Anal., 6 (1975), pp. 446–456.

[8] ———, *Topological dynamics of ordinary differential equations and Kurzweil equations*, J. Diff. Eqs., to appear.

[9] J. AUSLANDER AND P. SEIBERT, *Prolongations and generalized Liapunov functions*, Internat. Symp. on Nonlinear Differential Equations and Dynamical Systems, Academic Press, New York, 1963, pp. 454–462.

[10] ———, *Prolongations and stability in dynamical systems*, Ann. Inst. Fourier (Grenoble), 14 (1964), pp. 237–267.

[11] R. BELLMAN, *Introduction to Matrix Analysis*, McGraw-Hill, New York, 1960.

[12] ———, *Vector Lyapunov functions*, J. SIAM Control, 1 (1962), pp. 32–34.

[13] N. P. BHATIA AND O. HAJEK, *Local Semi-Dynamical Systems*, Springer-Verlag, Berlin, 1969.

[14] G. D. BIRKHOFF, *Quelques théorèmes sur le mouvement des systèmes dynamiques*, Bull. Soc. Math. France, 40 (1912), pp. 305–323. See also Birkhoff's *Collected Mathematical Papers*, Amer. Math. Soc., New York, 1950, pp. 654–672.

[15] ———, *Dynamical Systems*, Colloq. Publ. Vol. 9, Amer. Math. Soc., Providence, R. I., 1927. Revised edition, 1966.

[16] N. CHAFEE, *A stability analysis for a semilinear parabolic equation*, J. Diff. Eqs., 15 (1974), pp. 522–540.

[17] N. CHAFEE AND E. F. INFANTE, *A bifurcation problem for a nonlinear partial differential equation*, J. Appl. Anal., 4 (1974), pp. 17–37.

[18] M. A. CRUZ AND J. K. HALE, *Stability of functional differential equations of neutral type*, J. Diff. Eqs., 7 (1970), pp. 334–355.

[19] M. CUÉNOD AND A. DURING, *A Discrete-Time Approach for System Analysis*, Academic Press, New York, 1969.

[20] C. M. DAFERMOS, *An invariance principle for compact processes*, J. Diff. Eqs., 9 (1971), pp. 239–252.

[21] ———, *Uniform processes and semicontinuous Liapunov functionals*, Ibid., 11 (1972), pp. 401–415.

[22] ———, *Applications of the invariance principle for compact processes. I. Asymptotically dynamical systems*, Ibid., 9 (1971), pp. 291–299; *II. Asymptotic behavior of solutions of a hyperbolic conservation law*, Ibid., 11 (1972), pp. 416–424.

[23] ———, *Semiflows generated by compact and uniform processes*, Math. Systems Theory, 8 (1975), pp. 142–149.

[24] ———, *Contraction semigroups and trend to equilibrium in continuum mechanics*, to appear.

[25] C. M. DAFERMOS AND M. SLEMROD, *Asymptotic behavior of nonlinear contraction semigroups*, J. Functional Analysis, 13 (1973), pp. 97–106.
[26] F. DIPASQUANTONIO, *Stability in the first approximation and a critical case relating to nuclear reactor kinetics equations*, Nukleonik, 11 (1968), pp. 276–282.
[27] JOHN J. FEARNSIDES AND WILLIAM S. LEVINE, *On the determination of the asymptotic behavior of an inertially-oriented space station*, IEEE Trans. Automatic Control, AC-19 (1974), pp. 186–191.
[28] M. FIEDLER AND V. PTÁK, *On matrices with nonpositive off-diagonal terms and positive principal minors*, Czech. Math. J., 12 (1962), pp. 382–400.
[29] W. H. GOTTSHALK AND G. A. HEDLUND, *Topological Dynamics*, American Math. Soc., Providence, R. I., 1955.
[30] LJ. T. GRUJIĆ AND D. D. SILJAK, *Asymptotic stability and instability of large scale systems*, IEEE Trans. Automatic Control, AC-18 (1973), pp. 636–645.
[31] J. HADAMARD, *Sur les trajectoires en dynamique*, J. de Math., 3e Série, 3 (1897), pp. 331–387.
[32] W. HAHN, *Über die Anwendung der Methode von Ljapunov auf Differencegleichungen*, Math. Ann., 136 (1958), pp. 430–441.
[33] ———, *Theory and Application of Liapunov's Direct Method*, Prentice-Hall, Englewood Cliffs, N.J., 1963.
[34] A. HALANAY, *Quelques questions de la theórie de la stabilité pour les systèmes aux differences finies*, Arch. Rational. Mech. Anal., 12 (1963), pp. 150–154.
[35] ———, *Solutions périodiques et presque-périodiques des systèmes d'équations differences finies*, Ibid., 12 (1963), pp. 134–149.
[36] J. K. HALE, *A stability theorem for functional differential equations*, Proc. Nat. Acad. Sci. U.S.A., 50 (1963), pp. 942–946.
[37] ———, *Sufficient conditions for stability and instability of autonomous functional differential equations*, J. Diff. Eqs., 1 (1965), pp. 452–482.
[38] ———, *Dynamical systems and stability*, J. Math. Anal. Appl., 26 (1969), pp. 39–59.
[39] ———, *Ordinary Differential Equations*, Wiley–Interscience, New York, 1969.
[40] ———, *Functional Differential Equations*, Springer-Verlag, New York, 1971.
[41] J. K. HALE AND E. F. INFANTE, *Extended dynamical systems and stability theory*, Proc. Nat. Acad. Sci. U.S.A., 58 (1967), pp. 405–409.
[42] J. K. HALE AND J. P. LASALLE (eds.), *Differential Equations and Dynamical Systems*, Academic Press, New York, 1967.
[43] P. HARTMAN, *Ordinary Differential Equations*, John Wiley, New York, 1964.
[44] D. HENRY, *Geometric theory of semilinear parabolic equations*, unpublished manuscript, 1975.
[45] G. HORWICH AND P. A. SAMUELSON (eds.), *Trade, Stability, and Macroeconomics, Essays in Honor of Lloyd A. Metzler*, Academic Press, New York, 1974.
[46] J. HURT, *Some stability theorems for ordinary difference equations*, SIAM J. Numer. Anal., 4 (1967), pp. 582–596.
[47] E. F. INFANTE AND J. A. WALKER, *On the stability properties of an equation arising in reactor dynamics*, J. Math. Anal. Appl., to appear.
[48] E. I. JURY, *Inners and Stability of Dynamic Systems*, John Wiley, New York, 1974.
[49] R. E. KALMAN AND J. E. BERTRAM, *Control system analysis and design via the second method of Lyapunov. II. Discrete systems*, Trans. ASME Ser. D. J. Basic Engrg., 82 (1960), pp. 394–400.
[50] N. N. KRASOVSKII, *Stability of Motion with Applications of Lyapunov's Second Method to Differential Systems and Equations with Delay*, Stanford University Press, Stanford, Calif., 1963.
[51] C. KURATOWSKI, *Topology I*, Academic Press, New York, 1966.
[52] J. P. LASALLE, *The extent of asymptotic stability*, Proc. Nat. Acad. Sci. U.S.A., 46 (1960), pp. 363–365.
[53] ———, *Some extensions of Liapunov's second method*, IRE Trans. Circuit Theory, CT-7 (1960), pp. 520–527.

[54] J. P. LASALLE AND S. LEFSCHETZ, *Stability by Liapunov's Direct Method with Applications*, Academic Press, New York, 1961.

[55] J. P. LASALLE, *Asymptotic stability criteria*, Proc. Symp. Appl. Math. Hydrodynamic Instability, vol. 13, Amer. Math. Soc., Providence, R. I., 1962, pp. 299–307.

[56] ———, *Liapunov's second method*, Stability Problems of Solutions of Differential Equations, Proc. NATO Advanced Study Institute, Padua, Italy, Edizioni "Oderisi", Gubbio, 1966, pp. 95–106.

[57] ———, *An invariance principle in the theory of stability*, Differential Equations and Dynamical Systems, Proc. Internat. Symposium, Puerto Rico, Academic Press, New York, 1967, pp. 277–286.

[58] ———, *Stability theory for ordinary differential equations*, J. Diff. Eqs. 4 (1968), pp. 57–65.

[59] ———, *Stability theory and invariance principles*, Dynamical Systems. An International Symposium, Academic Press, New York, 1976.

[60] ———, *Vector Liapunov functions*, Bull. Instit. Math. Academia Sinica, Taiwan, 3 (1975), pp. 139–150.

[61] ———, *Stability of difference equations*, A Study in Ordinary Differential Equations (J. K. Hale, ed.) Studies in Mathematics Series, American Mathematical Association, to appear.

[62] J. P. LASALLE AND S. LEFSCHETZ, *Stability by Liapunov's Direct Method*, Academic Press, New York, 1961.

[63] J. P. LASALLE AND E. N. ONWUCHEKWA, *An invariance principle for vector Liapunov functions*, Dynamical Systems. An International Symposium, Academic Press, New York, 1976.

[64] S. LEFSCHETZ, *Stability of Nonlinear Control Systems*, Academic Press, New York, 1965.

[65] W. LEIGHTON, *On the construction of Liapunov functions for certain nonlinear differential equations*, Contrib. Diff. Eqs., 2 (1963), pp. 367–383.

[66] A. M. LIAPUNOV, *Problème général de la stabilité du mouvement*, Ann. of Math. Studies, Princeton Univ. Press, Princeton, N.J., 1947.

[67] A. I. LUR'E AND V. N. POSTNIKOV, *On the theory of stability of control systems*, Prikl. Mat. Mehk., 8 (1944), pp. 246–248.

[68] L. MARKUS, *Asymptotically autonomous differential systems*, Contributions to the Theory of Nonlinear Oscillations, vol. 3, Princeton Univ. Press, Princeton, N.J., 1956, pp. 17–29.

[69] J. C. MAXWELL, *On governors*, Proc. Roy. Soc. London, 16 (1868), pp. 270–283. A reprint of this paper and other selected papers can be found in Bellman and Kalaba, *Mathematical Trends in Control Theory*, Dover Publications, New York, 1964.

[70] O. MAYR, *Zur Frühgeshichte der Technischen Regelungen*, R. Oldenbourg Verlag, München, 1969; English transl.: *The Origins of Feedback Control*, M.I.T. Press, Cambridge, Mass., 1970.

[71] E. MCSHANE, *Integration*, Princeton Univ. Press, Princeton, N.J., 1947.

[72] L. A. METZLER, *Stability of multiple markets: the Hick's conditions*, Econometrica, 131 (1945), pp. 277–292.

[73] R. K. MILLER, *On almost periodic differential equations*, Bull. Amer. Math. Soc., 70 (1964), pp. 792–795.

[74] ———, *Asymptotic behavior of solutions of nonlinear differential equations*, Trans. Amer. Math. Soc., 115 (1965), pp. 400–416.

[75] ———, *Asymptotic behavior of nonlinear delay-differential equations*, J. Diff. Eqs., 1 (1965), pp. 293–305.

[76] ———, *Almost periodic differential equations as dynamical systems with applications to the existence of a.p. solutions*, Ibid., 1 (1965), pp. 337–345.

[77] R. K. MILLER AND G. R. SELL, *A note on Volterra integral equations and topological dynamics*, Bull. Amer. Math. Soc., 74 (1968), pp. 904–908.

[78] ———, *Existence, uniqueness and continuity of solutions of integral equations*, Ann. Mat. Pura Appl., 80 (1968), pp. 135–152.

[79] ———, *Volterra integral equations and topological dynamics*, Mem. Amer. Math. Soc., no. 102, 1970.

[80] ———, *Topological dynamics and its relations to integral equations and nonautonomous systems*, Dynamical Systems. An International Symposium, Academic Press, New York, 1976.

[81] W. E. MILNE, *Numerical Calculus*, Princeton Univ. Press, Princeton, N.J., 1949.

[82] L. M. MILNE-THOMPSON, *Calculus of Finite Differences*, Macmillan, London, 1933.

[83] A. P. MORGAN AND K. S. NARENDRA, *On the uniform asymptotic stability of certain linear nonautonomous differential equations*, Becton Center Tech. Rep. CT-64, Dept. of Engr. and Appl. Science, Yale University, New Haven, Conn., 1975.

[84] V. V. NEMYTSKII AND V. V. STEPANOV, *Qualitative Theory of Differential Equations*, Princeton Math. Ser. no. 22, Princeton Univ. Press, Princeton, N.J., 1960.

[85] C. OLECH, *On the global stability of an autonomous system in the plane*, Contrib. Diff. Eqs., 1 (1963), pp. 389–400.

[86] E. N. ONWUCHEKWA, *Stability of differential equations with applications to economics*, Doctoral dissertation, Brown University, Providence, R. I., 1975.

[87] Z. OPIAL, *Sur la dépendance des solutions d'un système d'équations différentielles de leur seconds membres. Application aux systèmes presque autonomes*, Ann. Polon. Math., 8 (1960), pp. 75–89.

[88] P. C. PARKS, *A stability criterion for a panel flutter problem via the second method of Liapunov*, Differential Equations and Dynamical Systems, Proc. Internat. Symp. Puerto Rico, Academic Press, New York, 1967, pp. 287–298.

[89] A. PAZY, *On the applicability of Lyapunov's theorem in Hilbert space*, SIAM J. Math. Anal., 3 (1972), pp. 291–294.

[90] T. K. L. PENG, *Invariance and stability for bounded uncertain systems*, SIAM J. Control, 10 (1972), pp. 679–690.

[91] O. PERRON, *Über Sabilität und asymptotische Verhalten der Lösungen eines Systems endlicher Differenzengleichungen*, J. Reine Angew. Math., 161 (1929), pp. 41–61.

[92] R. H. PLAUT, *Asymptotic stability and instability criteria for some elastic systems by Liapunov's direct method*, Quart. Appl. Math., (1972), pp. 535–540.

[93] E. J. PUTZER, *Avoiding the Jordan canonical form in the discussion of linear systems with constant coefficients*, Amer. Math. Monthly, 73 (1966), pp. 2–7.

[94] R. RESSIG, G. SANSONE AND R. CONTI, *Nonlinear Differential Equations of Higher Order*, Noordhoff, Leyden, 1974 (transl. of *Nichtlineare Differentialgleichungen höher Ordnung*, Edizioni Cremonese, Rome, 1969).

[95] H. H. SCHAEFER, *Banach Lattices and Positive Operators*, Springer-Verlag, New York, 1974.

[96] P. SEIBERT, *Stability under perturbations in generalized dynamical systems*, Internat. Symp. Nonlinear Differential Equations and Dynamical Systems, Academic Press, New York, 1963, pp. 463–473.

[97] G. R. SELL, *Nonautonomous differential equations and topological dynamics*, Trans. Amer. Math. Soc., 127 (1967), pp. 241–283.

[98] ———, *Lectures on Topological Dynamics and Differential Equations*, Van Nostrand–Reinhold, London, 1971.

[99] K. S. SIBIRSKY, *Introduction to Topological Dynamics*, Noordhoff, Leyden, Netherlands, 1975.

[100] V. SINGH, *A counterexample to harmonic linearization*, Proc. IEEE (letter), 63 (Nov. 1975), p. 1610.

[101] MARSHALL SLEMROD, *Asymptotic behavior of a class of abstract dynamical systems*, J. Diff. Eqs., 7 (1970), pp. 584–600.

[102] ———, *Asymptotic behavior of periodic dynamical systems on Banach spaces*, Ann. Mat. Bura Appl., 86 (1970), pp. 325–330.

[103] ———, *Nonexistence of oscillations in a nonlinear distributed network*, J. Math. Anal. Appl., 36 (1971), pp. 22–40.

[104] MARSHALL SLEMROD AND E. F. INFANTE, *An invariance principle for dynamical systems on Banach spaces*, Instability of Continuous Systems, Springer-Verlag, Berlin, 1971, pp. 215–221.

[105] A. Strauss and J. A. Yorke, *On asymptotically autonomous ordinary differential equations*, Math. System Theory, 1 (1967), pp. 175–182.

[106] D. R. Wakeman, *An application of topological dynamics to obtain a new invariance property for nonautonomous ordinary differential equations*, Doctoral dissertation, Brown University, Providence, R. I., 1973.

[107] J. A. Walker, *Liapunov analysis of the generalizer Pflueger problem*, ASME J. Appl. Mech., 39 (1972), pp. 935–938.

[108] ———, *Energy-like Liapunov functionals for linear elastic systems in a Hilbert space*, Quart. Appl. Math., 30 (1973), pp. 465–480.

[109] ———, *On the application of Liapunov's direct method to linear dynamical systems*, J. Math. Anal. Appl., to appear.

[110] T. Yoshizawa,, *Asymptotic behavior of solutions of a system of differential equations*, Contrib. Diff. Eqs., 1 (1963), pp. 371–387.

[111] ———, *Stability Theory by Liapunov's Second Method*, Publication no. 9, Math. Soc. Japan, Tokyo, 1966.

[112] V. I. Zubov, *Methods of A. M. Lyapunov and Their Application*, Noordhoff, Groningen, Netherlands, 1964.

# APPENDIX A

## Limiting Equations and Stability of Nonautonomous Ordinary Differential Equations

*ZVI ARTSTEIN*†

**1. A key idea.** We are interested in systems that are governed by a nonautonomous, i.e., time-dependent, ordinary differential equation

$$(*) \qquad \dot{x} = f(x, s).$$

The relation (*) describes the law of motion a solution $x(s)$ has to observe. We shall investigate some aspects of the relation between the asymptotic behavior of solutions of (*) and the changes in the law of motion while time progresses.

The underlying motivation of the investigation can be summarized by the following.

*Idea.* Let us use the asymptotic behavior of the time-dependent function $f(x, s)$ in the investigation of the asymptotic behavior of solutions of $\dot{x} = f(x, s)$.

As natural and simple as this idea looks, it is only recently that techniques exploiting it have been developed and used in relation to stability and asymptotic behavior of solutions of nonlinear equations. It is the purpose of this paper to demonstrate and explain one method inspired by the displayed idea, namely, the use of the limiting equations (to be defined in a moment) of equation (*). An historical account and some related subjects will be given in a separate section (see § 16).

One particular situation where the idea was extensively used is the asymptotically autonomous case

$$(1.1) \qquad \dot{x} = g(x) + h(x, s),$$

where the perturbation $h(x, s)$ tends to zero (in a certain sense) as $s \to \infty$. Here—at least intuitively—it is clear that the limiting behavior of the time-dependent law (1.1) is portrayed by the time-independent equation $\dot{x} = g(x)$. We say then that the latter is a limiting equation of (1.1), and we even may say that it is *the* limiting equation since a unique limiting equation exists. In the general case an equation might have more than one limiting equation.

---

* Invited address at the NSF–CBMS Conference on the Stability of Dynamical Systems, Theory and Applications, Mississippi State University, August 1975.

† Lefschetz Center for Dynamical Systems, Division of Applied Mathematics, Brown University, Providence, R. I. 02912. The research was supported by the National Science Foundation under GP 28931X3 and by the Office of Naval Research under NONR N1467-AD-101000907.

We want to clarify somewhat the vague phrase "asymptotic behavior of the equation $\dot{x}=f(x,s)$." The asymptotically autonomous case (1.1) is clear. For $t_0$ large the initial value problem (1.1) with $x(t_0)=x_0$ is close to the initial value problem $\dot{x}=g(x)$ and $x(t_0)=x_0$. In the general case we want to trace the behavior of $\dot{x}=f(x,s)$, $x(t_k)=x_0$ for sequences $t_k\to\infty$. Hopefully for certain sequences the limiting behavior will be described by an equation $\dot{x}=g(x,s)$. Here we encounter a formal problem which does not appear in the asymptotically autonomous case. If we were to compare the behavior of $\dot{x}=f(x,s)x(t_k)=x_0$ to $\dot{x}=g(x,s)$, $x(t_k)=x_0$, we might miss the whole point of the idea, since the behavior of $g(x,s)$ itself changes in time. Therefore a machinery has to be developed which will enable us to compare the initial value problem for different initial times to a fixed behavior. The concept of a *translate* of an equation will do the job.

DEFINITION A. The *translate* by $t$ of the function $f(x,s)$ is the function $f^t$ defined by $f^t(x,s)=f(x,t+s)$.

Notice that the equation $\dot{x}=f^t(x,s)$ represents a change in the time variable with respect to $\dot{x}=f(x,s)$, namely, $s\to s-t$. Solutions of $\dot{x}=f(x,s)$, $x(t)=x_0$ are identical up to this change in time with solutions of $\dot{x}=f^t(x,s)$, $x(0)=x_0$. This enables us to compare solutions and equations on different domains and different initial times, and consequently discuss the limiting behavior of $f(x,s)$ for $s$ large.

DEFINITION B. The equation $\dot{x}=g(x,s)$ is a *limiting equation of* $\dot{x}=f(x,s)$ if there exists a sequence $t_k\to\infty$ so that the translates $f^{t_k}$ converge to $g$ as $k\to\infty$.

Notice that the definition is not complete unless we specify the meaning of convergence. In general, we can use several types of convergence in order to achieve different goals. Clearly whether $\dot{x}=g(x,s)$ is or is not a limiting equation depends on the type of convergence adopted.

The following definition and notation are natural, but notice that again the outcome depends on the type of convergence used in the construction of the limiting equations.

DEFINITION C. The equation $\dot{x}=f(x,s)$ is positively precompact if whenever $t_k\to\infty$ a subsequence $\rho_i$ exists such that $f^{\rho_i}$ converges.

**Notation.** The collection of the limiting equations of $\dot{x}=f(x,s)$ will be denoted by $L^+(f)$.

The discussion and definitions above are related to the behavior of the equation and solutions as $s\to\infty$. There is no problem in extending the definitions to the case $s\to-\infty$. The reader will probably not be surprised if we then use the terminology negatively precompact and the notation $L^-(f)$.

Naturally, almost automatically, while talking about a limiting equation we had in mind an ordinary differential equation. Let us display the following idea now, although we shall not treat it rigorously until later (starting in § 11) in the paper.

*Idea.* The asymptotic behavior of $\dot{x}=f(x,s)$ and its solutions might not be described by an ordinary differential equation, but we may try to use other equations ("unordinary") as limiting equations.

**2. Invariance, limiting equations and continuous dependence.** We are still at an intuitive level. We want to focus on a particular application of the limiting equations, namely, invariance properties. This will hopefully give us a supporting view for introducing the limiting equations.

Let $\boldsymbol{x} = x(s)$ be a solution of the ordinary differential equation

$$(*) \qquad \dot{x} = f(x, s).$$

The $\omega$-limit set of $\boldsymbol{x}$, denoted $\Omega(\boldsymbol{x})$, consists of all points $y$ such that a sequence $t_k \to \infty$ exists with $x(t_k) \to y$. In case (*) is autonomous, i.e., has the form

$$(2.1) \qquad \dot{x} = f(x),$$

it is well known that $\Omega(\boldsymbol{x})$ has an invariance property with respect to (2.1). If the solution of the initial value problem related to (2.1) is unique, then the invariance is as follows. For any $y \in \Omega(\boldsymbol{x})$ the solution of (2.1) through $y$ stays in $\Omega(\boldsymbol{x})$ whenever it is defined. The proof is based essentially on the continuous dependence property of (2.1). Given $y \in \Omega(\boldsymbol{x})$ we pick a sequence $x(t_k) \to y$ (see Fig. 1(a)). Continuous dependence on initial data says roughly that "the limit of the solutions through $x(t_k)$ is the solution through the limit point $y = \lim x(t_k)$." We know, by definition, that the limit of the solutions is in $\Omega(\boldsymbol{x})$, so the solution through $y$ stays in $\Omega(\boldsymbol{x})$.

In trying to formulate the analogous invariance result for the time-dependent equation (*) we encounter several problems. The serious one is to identify an equation with respect to which $\Omega(\boldsymbol{x})$ will be invariant. There is no hope, of course, that solutions of (*) through $y \in \Omega(\boldsymbol{x})$ will stay, in general, in $\Omega(\boldsymbol{x})$. We shall now explain how the limiting equations come into the picture.

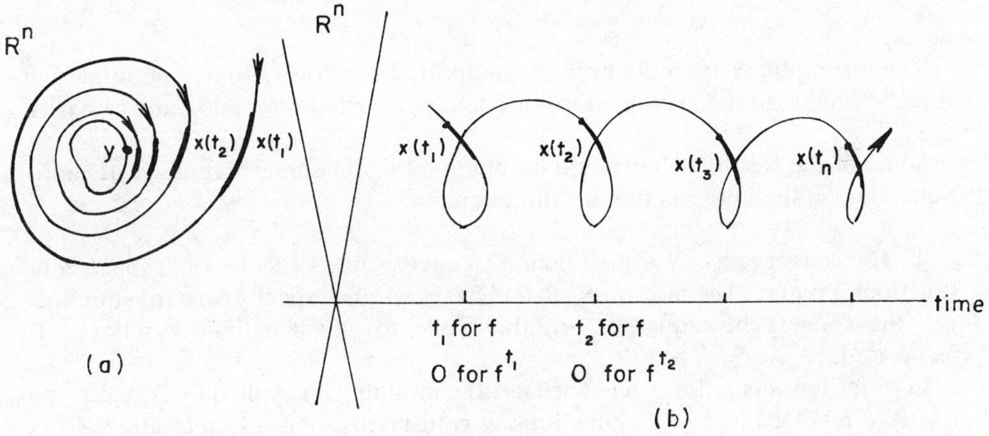


FIG. 1

Given $y \in \Omega(\boldsymbol{x})$ we pick a sequence $x(t_k) \to y$, (see again Fig. 1(a)). But we have to take into consideration the changes in $f(x, s)$ while time progresses, so we look at the state-time chart (Fig. 1(b)) of the solution, and to each point $t_k$ we associate the translate $f^{t_k}$. Suppose that $f^{t_k}$ (or a subsequence) converges to $g$, and suppose that the convergence is such that the following continuous dependence result holds: "The limit of the solutions of $\dot{x}=f^{t_k}(x, s)$ through $(x(t_k), 0)$ is the solution of $\dot{x}=g(x, s)$ through the limit $(y, 0)$." By recalling that the solution of $\dot{x}=f^{t_k}(x, s)$ through $(x(t_k), 0)$ is a translate of the original solution $\boldsymbol{x}$, we conclude that the solution of $\dot{x}=g(x, s)$, $x(0)=y$ stays in $\Omega(\boldsymbol{x})$. To sum up, the limiting equations are those with respect to which $\Omega(\boldsymbol{x})$ has a certain invariance property.

Notice that the analysis above also suggests the type of convergence appropriate for handling invariance properties. We need a convergence which guarantees continuous dependence of the solutions on the initial data *and* the right-hand side of the equation.

**3. The assumptions.** Starting now with the formal part of the paper, we place here our assumptions and restrictions on the differential equation

$$\dot{x}=f(x, s). \tag{*}$$

Our state space is $R^n$, the $n$-dimensional Euclidean space. The norm of $x \in R^n$ will be denoted by $|x|$. We assume that $f$ is continuous in $x$, measurable in $s$, satisfies the Carathéory conditions locally (see [15, p. 28]) and uniformly satisfies:

ASSUMPTION (A). For every compact $K \subset R^n$ there is a nondecreasing function $\mu_K:[0, \infty) \to [0, \infty]$, continuous at 0 with $\mu_K(0)=0$ and such that, whenever $u:[a, b] \to K$ is continuous, the integral $\int_a^b f(u(s), s)\, ds$ is well defined and the estimate

$$\left|\int_a^b f(u(s), s)\, ds\right| \leqq \mu_K(b-a)$$

holds.

The assumptions are quite mild. Assumption (A) allows $f$ to be unbounded in time, although on the average it is bounded. For further discussion and remarks, see [5].

Some of the results below could be obtained under somewhat eased assumptions, but we shall not pursue this direction.

**4. The convergence.** We shall define a convergence structure on a space $\mathcal{G}$ of functions $g=g(x, s)$ defined on $R^n \times R$. That is, we shall specify certain sequences $g_k$—the convergent sequences—in the space to which a limit $g_0=\lim g_k$ is associated.

In order to specify the space $\mathcal{G}$ we use the moduli $\mu_K$ supplied by Assumption (A). We let $\mathcal{G}$ consist of all functions $g$, continuous in $x$, and measurable in $s$ satisfying

$$\left|\int_a^b g(u(s), s)\, ds\right| \leqq \mu_K(b-a) \tag{4.1}$$

whenever $u:[a, b] \to K$ is continuous and $K \subset R^n$ compact.

Clearly all the translates $f^t$ of $f$ (and recall that $f^t(x, s)=f(x, t+s)$) belong to $\mathcal{G}$.

DEFINITION 4.2. The sequence $g_k$ in $\mathcal{G}$ converges to $g_0$ if whenever $u_k$ is a sequence of continuous functions on $[a, b]$ which converges uniformly to $u_0$, then

$$\int_a^b g_k(u_k(s), s)\, ds \to \int_a^b g_0(u_0(s), s)\, ds. \tag{4.3}$$

*Remark.* Condition (4.1) implies that the convergence in (4.3) is uniform in $b$ on compact sets.

In case $g_k$ are translates $f^{t_k}$ of $f$ the convergence (4.3) can be rewritten as

$$\int_a^b f(u_k(s), t_k + s)\, ds \to \int_a^b g_0(u_0(s), s)\, ds.$$

Another property that will be useful in the sequel is that if $f^{t_k}$ converges to $g$, then $f^{\tau+t_k}$ converges to the translate $g^\tau$. In referring to the limiting equations this means that $L^+(f)$ is closed under translation.

We want to investigate a little bit the structure of the space $\mathcal{G}$ with the convergence given in Definition 4.2, although this structure will hardly be used in most of the analysis below. It is easily checked that the convergence satisfies properties (i) through (iii) of [17, p. 188], and thus the space $\mathcal{G}$ with the convergence is eligible to the title $\mathcal{L}^*$-space (sometimes called a convergence space). Some questions are in order now:

1. Is the convergence generated by a topology or a metric?
2. Can we supply a better representation for the convergence?
3. How is the convergence related to other structures used already in the literature?
4. Can it be relaxed?

Dealing in detail with some answers to these questions is too much off our main course, so we postpone the discussion until § 15.

**5. Some examples and remarks.**

**5.1.** The equation

$$\dot{x} = g(x) + h(x, s)$$

is asymptotically autonomous with respect to the convergence in Definition 4.2 if whenever the sequence $u_k$ of continuous functions on $[a, b]$ converges uniformly to $u_0$ and $t_k \to \infty$, then

$$\int_a^b h(u_k(s), t_k + s)\, ds \to 0. \tag{5.1}$$

This type of convergence is weaker than the one used by Strauss and Yorke [40] or Markus [26] (cf. also [5]).

**5.2.** If $h(x, s)$ satisfies (5.1) above, then the equation $\dot{x} = f(x, s)$ and the perturbed equation $\dot{x} = f(x, s) + h(x, s)$ share the same family of limiting equations.

**5.3.** The limiting equations of

$$\dot{x} = \sin (s)^{1/2}$$

are exactly all the equations $\dot{x} = \alpha$ where $\alpha$ is a constant and $-1 \leqq \alpha \leqq 1$.

**5.4.** The equation

$$\dot{x} = \sin (s)^2$$

has only $\dot{x} = 0$ as a limiting equation. More general, in $R^2$, the equation

$$\dot{x} = (e^{\tau} \sin e^{2\tau}, e^{\tau} \sin e^{2\tau})$$

has $\dot{x} = 0$ as a limiting equation, although the norm of the right-hand side converges to infinity as $\tau \to \infty$ (see Strauss and Yorke, [40]).

**5.5.** The limiting equations of the periodic equation $\dot{x} = p(x, s)$ with $p(x, s+T) = p(x, s)$ form a cycle, $p^t$ for $0 \leqq t \leqq T$, in the function space.

**5.6.** If $r(\tau) \to 1$ slowly enough as $\tau \to \infty$, then the limiting equations of

$$\dot{x} = \sin r(\tau)\tau$$

are in the periodic circle $\dot{x} = \sin (\alpha + \tau)$ for $0 \leqq \alpha \leqq 2\pi$, and are translates of one periodic equation. Nevertheless, the equation itself is *not* a perturbation, by a term which converges to zero, of a periodic equation.

**5.7.** Let $f(x, s)$ on $R \times R$ satisfy: if, for some positive integer $k$, $s - x = 2^k$, then $f(x, s) = 1$; if $|s - x - 2^k| \leqq 2^{-k}$, $0 \leqq f(x, s) \leqq 1$; otherwise, $f(x, s) = 0$. Then $\dot{x} = f(x, s)$ is not positively precompact. But it has one limiting equation, which is $\dot{x} = 0$.

**5.8.** If an equation has a unique limiting equation, the latter has to be autonomous, since it is equal to all its translates (which also are limiting equations).

**5.9.** If an equation is positively precompact and has a unique limiting equation, it is necessarily of the form

$$\dot{x} = g(x) + h(x, s),$$

where $h^t \to 0$, as $t \to \infty$, i.e., asymptotically autonomous. This is easily verified. Without the positive precompactness the conclusion is not true, as Example 5.7 shows.

**6. The continuous dependence.** The purpose of this section is to show that the solutions of $\dot{x} = g(x, s)$, $x(0) = x_0$ depend continuously on $x_0 \in R^n$ and $g \in \mathcal{G}$, where the latter is endowed with the convergence given in Definition 4.2. We do not assume uniqueness of solutions so we shall use a formulation of continuous dependence due to Kamke (see [39, p. 46] or [16, Theorem 3.2, p. 14]).

THEOREM 6.1. *Let $y_k \to y_0$ in $R^n$ and $g_k \to g_0$ in $\mathcal{G}$. For each $k = 1, 2, \cdots$, let $\mathbf{x}_k = x_k(s)$ be a maximally defined solution of $\dot{x} = g_k(x, s)$, $x(0) = y_k$. Then a subsequence $\mathbf{x}_m$ exists which converges to a maximally defined solution $\mathbf{x}_0$ of $\dot{x} = g_0(x, s)$, $x(0) = y_0$, and the convergence is uniform on compact subintervals of the domain of $\mathbf{x}_0$.*

In different wording the result was proved in [5, Theorem 5.3]. The sketch of the proof in [3, Theorem 3.1] is valid with only minor changes to the present situation.

The continuous dependence would be stated more elegantly in terms of the set $s(y, g)$ of maximally defined solutions of $\dot{x} = g(x, s)$, $x(0) = y$. A metric structure that represents the uniform convergence on compact intervals can be associated with the collection of solutions (although their domains are different) (see [5, § 4]). Then Theorem 5.1 simply means that $s(y, g)$ is compact-valued and upper semicontinuous in the pair $(y, g)$ (see [2]).

**7. Invariance properties and invariance principles.** We shall formulate and rigorously prove the results sketched in § 2. The general invariance properties will follow. Again, we deal with

$$(*) \qquad \dot{x} = f(x, s).$$

For convenience we recall some definitions. The translate $f^t$ of $f$ is defined by $f^t(x, s) = f(x, t+s)$. A limiting equation of (*) is an equation $\dot{x} = g(x, s)$ such that for a certain sequence $t_k \to \infty$ the convergence $f^{t_k} \to g$ holds. Here the convergence is that of Definition 4.2. The equation (*) is positively precompact if $t_k \to \infty$ implies that a subsequence of $f^{t_k}$ converges. The family of limiting equations of (*) is denoted by $L^+(f)$. The $\omega$-limit set of a function $\mathbf{x} = x(s)$ is the set of limits $\lim x(t_k)$ for sequences $t_k \to \infty$; it is denoted by $\Omega(\mathbf{x})$.

THEOREM 7.1. *Let $\mathbf{x} = x(s)$ be a solution of (*) defined for $t_0 \leqq s < \infty$. If for a sequence $t_k \to \infty$ the vectors $x(t_k) \to y_0$ in $R^n$, and the functions $f^{t_k} \to g$ in $\mathcal{G}$, then a maximally defined solution $\mathbf{y} = y(s)$ of $\dot{x} = g(x, s)$, $x(0) = y_0$ exists such that $y(s) \in \Omega(\mathbf{x})$ for every $s$ in its domain.*

*Proof.* Let $\mathbf{x}_k = x_k(s)$ be defined by $x_k(s) = x(t_k + s)$. Then $\mathbf{x}_k$ is a solution of $\dot{x} = f^{t_k}(x, s)$, $x(0) = x(t_k)$. By Theorem 6.1 a subsequence $\mathbf{x}_m$ exists which converges to a solution $\mathbf{y}$ of $\dot{x} = g(x, s)$, $x(0) = y_0$. Let $\tau$ be in the domain of $\mathbf{y}$. Then $x(t_k + \tau) = x_k(\tau) \to y(\tau)$ as $k \to \infty$. This means that $y(\tau) \in \Omega(\mathbf{x})$. This completes the proof.

Several invariance properties of the $\omega$-limit set can be deduced from Theorem 7.1. We shall state some.

THEOREM 7.2. (local-semi-quasi invariance property). *Let $\mathbf{x}$ be a solution of (*). If (*) is positively precompact, then for every $y_0 \in \Omega(\mathbf{x})$ there is a limiting equation $\dot{x} = g(x, s)$ of (*) and there exists a solution $\mathbf{y} = y(s)$ of $\dot{x} = g(x, s)$, $x(0) = y_0$, which stays in $\Omega(\mathbf{x})$ on its entire domain.*

*Proof.* Pick a sequence $x(t_k) \to y_0$. Then a converging subsequence of $f^{t_k}$ exists, with limit $g$. Now use Theorem 7.1.

*Explanation and modifications.* The "local" in the title of Theorem 7.2 indicates that the solution through $y_0$ might not be defined for all $s$. A well-known example is illustrated in Fig. 2, where the equation on the line $y=1$ is, say, $\dot{x}=1+x^2$, $\dot{y}=0$. Here the $\omega$-limit set is unbounded. If $\Omega(\boldsymbol{x})$ is compact, then $\boldsymbol{y}$ is defined for all $s\in R$, since a solution with a bounded maximal domain is unbounded. In this case (boundedness of $\Omega(\boldsymbol{x})$) the "local" can be dropped.

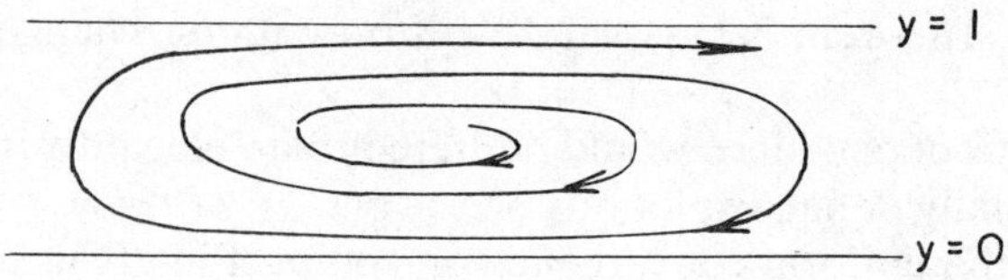


FIG. 2

The "semi" and "quasi" in Theorem 7.2. are to remind us of the two existential quantifiers in the statement. Indeed, not every limiting equation will suit, and the one that does might have other solutions through $(y_0, 0)$ which do not stay in $\Omega(\boldsymbol{x})$. In case solutions of the initial value problem for the limiting equations are unique the "quasi" can be dropped. If there is a unique limiting equation (and together with positive precompactness it means that the equation is asymptotically autonomous), we can get rid of the "semi".

THEOREM 7.3 (another semi-quasi invariance property). *Let $\boldsymbol{x}=x(s)$ be a solution of* (*) *and assume $\Omega(\boldsymbol{x})$ is nonempty and compact. Then for every limiting equation $\dot{x}=g(x,s)$ of* (*) *there exists a vector $y_0\in\Omega(\boldsymbol{x})$ such that a solution $\boldsymbol{y}=y(s)$ of $\dot{x}=g(x,s)$, $x(0)=y_0$ exists, with $y(s)\in\Omega(\boldsymbol{x})$ for all $s\in R$.*

The proof is similar to that of the previous theorem. We only have to replace the order of constructing the sequences. We first pick $f^{t_k}\to g$. The compactness of $\Omega(\boldsymbol{x})$ implies the compactness of $\{x(s): s\geqq t_0\}$ and therefore a subsequence $x(t_m)\to y_0\in\Omega(\boldsymbol{x})$ exists. Now we apply Theorem 7.1 to the sequence $t_m$. The compactness of $\Omega(\boldsymbol{x})$ implies that $\boldsymbol{y}$ is defined on the whole line $R$.

COROLLARY 7.4. *If $x(s)$ converges to a point $y_0$ as $s\to\infty$, then $y_0$ is a rest point for any limiting equation.*

The invariance properties of the $\omega$-limit set together with techniques involving Liapunov's direct method form a very powerful tool in detecting stability and asymptotic stability and locating regions of stability and asymptotic stability for nonlinear systems. This powerful tool is the LaSalle invariance principle. The basic idea is to use direct methods to locate regions of stability and attractivity of a set $E$ and then to refine the result by using invariance properties of subsets of $E$. In many examples this invariance principle gives strictly sharper results comparing with Liapunov theory. (Some references are [9], [20]–[24], [33], [34], [35], [41]. See also the remarks in § 16 below.)

We shall demonstrate now how the second part of this program works. The next section will be devoted to the first half, i.e., the problem of how to locate $E$.

DEFINITION 7.5. A set $Q \subset R^n$ is *local-semi-quasi invariant with respect to a collection of equations* $\mathscr{L} = \{\dot{x} = g(x, s)\}$ if for every $y_0 \in Q$ there is an equation $\dot{x} = g(x, s)$ in $\mathscr{L}$ and a maximally defined solution $\mathbf{y} = y(s)$ through $(y_0, 0)$ of $\dot{x} = g(x, s)$ such that $y(s) \in Q$ for every $s$ in its domain.

The explanation and the modifications following Theorem 7.2 apply here as well.

THEOREM 7.6. *Suppose* $\mathbf{x} = x(s)$ *is a bounded solution of* (*) *and that* $x(s)$ *converges to a set* $E \subset R^n$ *as* $s \to \infty$. *If* (*) *is positively precompact, then* $x(s)$ *converges to the largest set M in E which is local-semi-quasi invariant with respect to* $L^+(f)$.

The proof is an immediate consequence of Definition 7.5 and Theorem 7.2.

*Example* 7.7. In the $(x, y) \in R^n \times R^m$ space consider the equation

$$\dot{x} = f_1(x, y, s),$$

$$\dot{y} = f_2(x, y, s)y + f_3(x, y).$$

Assume that $f_3(x, y) \neq 0$ if $x \neq 0$. Also assume that the equation is positively precompact. If a bounded solution $(x(s), y(s))$ satisfies $y(s) \to 0$, then $x(s) \to 0$ as well. Indeed, the $\omega$-limit set has $y$-coordinates zero and we may set $E = \{(x, y) : y = 0\}$. But the only invariant set in $E$ is the origin, since for any limiting equation $\dot{y} = f_3(x, y)$ on $E$. (The example is a paraphrase, a restricted one, of Levin [25].)

A result similar to Theorem 7.6 can be formulated also with respect to the invariance property given in Theorem 7.3. We shall now demonstrate how to use it.

*Example* 7.8. The equation is the same as in Example 7.7, but we drop the assumption that the equation is positively precompact. Instead we assume that at least one limiting equation exists. The conclusion is that a bounded solution $(x(s), y(s))$ which satisfies $y(s) \to 0$ has the origin in its $\omega$-limit set. If in addition, the origin is uniformly stable, then $x(s) \to 0$ too. The proof is an immediate consequence of Theorem 7.6.

We want to conclude this section by referring to the work of Infante and the author [7]. Here a quantitative result, the rate of growth of a certain damping coefficient, has been obtained from qualitative considerations similar to those of the present section. The result is an application of Corollary 7.4 above. Section 4 in [7] explains the relation to the invariance principle.

**8. How to locate *E*.** We shall discuss now the first half of the invariance principle, namely how, with the aid of Liapunov functions, a set $E$ can be found toward which solutions of

$$(*) \qquad \dot{x} = f(x, s)$$

converge.

A function $V(x, s)$ is a *Liapunov function with respect to* (*) if $V$ is continuous, $V(x, s) \geqq 0$ and $V(\mathbf{x}(s), s)$ is a nonincreasing function of $s$ for every solution $\mathbf{x}(s)$

of (*). We then define

$$V'(x, s) = \limsup_{h \to 0^+} \frac{1}{h}[V(\boldsymbol{x}(s+h), s+h) - V(\boldsymbol{x}(s), s)],$$

where the lim sup is also over all solutions $\boldsymbol{x}(s)$ through $(x, s)$. Under quite mild assumptions $V'$ can be computed directly from the equation (*), without using the solutions at all (see LaSalle [23] and Yoshizawa [43]).

The following result is a generalization of Yoshizawa [42] given by LaSalle [23]. In [23] it is assumed that $f(x, s)$ is bounded in $s$ for $x$ bounded, but it is easy to check that the proof extends to the present situation.

THEOREM 8.1. *Suppose $f(x, s)$ satisfies Assumption* (A). *Let $V$ be a Liapunov function for* (*) *and suppose $V'(x, s) \leqq W(x) \leqq 0$, where $W$ is a continuous function. Denote $E = \{x : W(x) = 0\}$. Then any bounded solution of* (*) *converges to $E$ as time goes to infinity.*

Extensions of Theorem 8.1 were obtained by Burton [8] and Haddock [13], [14], mainly in the direction of easing the requirement $V'(x, s) \leqq W(x)$ and the boundedness of $f$. (The estimation $V'(x, s) \leqq W(x)$ means that an autonomous function $W$ dominates the rate in which the nonautonomous $V$ decreases, which seems to be a restrictive assumption.) The following result is basically the one announced in Haddock [14, Theorem 3].

THEOREM 8.2. *Let $H \subset R^k$ be a closed set. Suppose that for every compact set $K$ disjoint from $H$ there is a $\delta > 0$ such that*

$$V'(x, s) \leqq -\delta |f(x, s)| + e(s)$$

*for $x \in K$, where $e(s)$ is integrable. Then a bounded solution of* (*) *either converges to a constant or converges to $H$ as time tends to infinity.*

The reader surely noticed that the sets $E$ and $H$ in Theorems 8.1 and 8.2 were located by using Liapunov functions of the original equation (*). In the context of our paper it is only natural to note the following.

*Idea.* Can we use Liapunov functions of the limiting equations of (*) to locate the set $E$?

An affirmative answer will be of advantage, since in many cases the limiting equations have simpler structure, and are more easily handled, for example, the asymptotically autonomous case. The author has done some work with respect to this idea, and it hopefully will appear in [6]. We wish to report here some partial results.

Difficulties arise even in the asymptotically autonomous case $\dot{x} = g(x) + h(x, s)$. If $V(x)$ is a Liapunov function for $\dot{x} = g(x)$, then a solution of the perturbed equation does not necessarily converge to $E = \{x : V'(x) = 0\}$. An example is illustrated in Fig. 3. Here $g(x) = 0$ on the segment AB. The trajectories of $\dot{x} = g(x)$ are the solid curves while the dotted curves represent level curves of $V$. Then $V'(x) = 0$ exactly for $x$ in the segment AB, and indeed every solution of $\dot{x} = g(x)$ converges to $E$. But with a small perturbation $h$ a solution of $\dot{x} = g(x) + h(x, s)$ might "climb" from B to A (climb with respect to $V$) and go back along a solution of $\dot{x} = g(x)$. The outcome might be that the fat closed curved is the $\omega$-limit set.

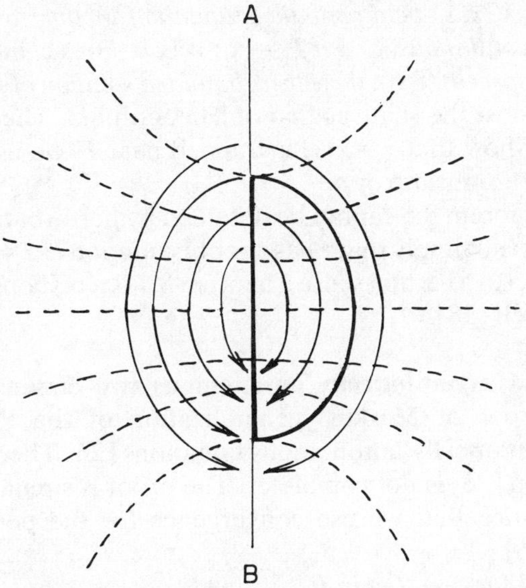


FIG. 3

What makes the example above work is the instability of $E=\{x: V'(x)=0\}$. If $E$ is stable we have the following.

THEOREM 8.3. *Suppose that*

$$\dot{x}=g(x)+h(x, s) \tag{8.4}$$

*is asymptotically autonomous, (i.e., $h^t \to 0$ as $t\to\infty$). Let $V=V(x)$ be a Liapunov function for $\dot{x}=g(x)$. If $E=\{x: V'(x)=0\}$ is compact, and stable with respect to $\dot{x}=g(x)$, then every bounded solution $\boldsymbol{x}=x(s)$ of (8.4) converges to E.*

*Proof.* The $\omega$-limit set $\Omega(\boldsymbol{x})$ is invariant under $\dot{x}=g(x)$. Let $y_0\in\Omega(\boldsymbol{x})$ and let $\boldsymbol{y}=y(s)$ be a solution of $\dot{x}=g(x)$ which stays in $\Omega(\boldsymbol{x})$. Then $\Omega(\boldsymbol{y})\subset\Omega(\boldsymbol{x})$ and also $\Omega(\boldsymbol{y})\subset E$. Therefore, $x(s)$ comes close to $E$ infinitely many times. It is not hard to show now that if $x(s)$ is close to $E$ for a certain large $s$, then it is "trapped" in a neighborhood of $E$. This follows from the asymptotic stability of $E$ and from the fact that for $t$ large the equation $\dot{x}=g(x)+h^t(x, s)$ is close to $\dot{x}=g(x)$.

The proof of Theorem 8.3 could be based on perturbation arguments, but notice that we only used the fact that for $t$ large $g+h^t$ is close to $g$. This method generalizes to systems that are not asymptotically autonomous, where perturbation theory does not apply. For instance,

THEOREM 8.5. *Suppose that (*) is positively precompact and that the convergence of $f^t$ to $L^+(f)$ is induced by a metric. Suppose that $L^+(f)$ is a cycle in the function space $\mathscr{G}$ which is generated by a periodic equation*

$$\dot{x}=p(x, s) \tag{8.6}$$

*with period T. Let* $V(x, s)$ *be a periodic Liapunov function for* (8.6). *If* $V'(x, s) \leqq W(x) \leqq 0$ *with W continuous, and if* $E = \{x : W(x) = 0\}$ *is compact and asymptotically stable with respect to* (8.6), *then every bounded solution of* (*) *converges to E.*

The proof is almost the same as that of Theorem 8.3. The only change is that when we want to show that $\boldsymbol{x} = x(s)$ is trapped near $E$ we use the fact that for $t$ large $f^t$ is close to a translate of $p$.

The result in Theorem 8.5 cannot be obtained by perturbation arguments since the assumptions do not imply that $f$ has a representation as $f = p + h$, where $h^t \to 0$ (see Example 5.7). (In a certain sense Theorem 8.7 is a response to the concluding remark of § 6 in Sell [38].)

**9. A remark on asymptotically autonomous two-dimensional systems.** We shall sketch a proof of Markus' generalization of the Poincaré–Bendixson theorem for asymptotically autonomous equations [26, Theorems 7]. (Unfortunately, the proof in [26] is not complete.) The proof is similar in spirit to that of Theorem 8.3. Notice that we use convergence for the perturbation which is weaker than in [26].

THEOREM 9.1 (Markus). *Let in* $R^2$

$$\dot{x} = g(x) + h(x, s) \tag{9.2}$$

*be asymptotically autonomous, (i.e.,* $h^t \to 0$ *as* $t \to \infty$*). Let a solution* $\boldsymbol{x} = x(s)$ *of* (9.2) *lie in a compact set* $K \subset R^2$ *and suppose* $\Omega(\boldsymbol{x})$ *does not contain any critical points of* $\dot{x} = g(x)$. *Also assume that* $\dot{x} = g(x)$ *has the uniqueness property. Then* $\Omega(\boldsymbol{x})$ *is the union of closed orbits of* $\dot{x} = g(x)$.

*Proof.* Let $y_0 \in \Omega(\boldsymbol{x})$. If $y_0$ lies on a closed orbit of $\dot{x} = g(x)$ we are finished. If not, the Poincaré–Bendixson theorem (see [15]) implies that the solution $\boldsymbol{y}$ of $\dot{x} = g(x)$ through $y_0$ spirals into (out-to) a closed orbit (since $\Omega(\boldsymbol{y}) \subset \Omega(\boldsymbol{x})$ there are no critical points in $\Omega(\boldsymbol{y})$). Also, $\Omega(\boldsymbol{y})$ is then asymptotically stable from the outside (inside). Since $\Omega(\boldsymbol{y}) \subset \Omega(\boldsymbol{x})$ it follows that $x(s)$ comes close to $\Omega(\boldsymbol{y})$ infinitely many time as $s \to \infty$, but for $s$ large it is "trapped" near $\Omega(\boldsymbol{y})$. So $y_0$ cannot be in $\Omega(\boldsymbol{x})$.

**10. Positive precompactness in the restricted sense.** On several occasions throughout the paper, we have used the condition that

$$\dot{x} = f(x, s) \tag{*}$$

is positively precompact (see Definition C). See Theorems 7.2, 7.6 and Example 7.7. It is more than appropriate to supply conditions which guarantee it. The "restricted" in the title is related to the idea which concludes § 1. Later in the paper (§ 14) we shall examine positive precompactness where the limiting equations might not be ordinary differential equations. Then the conditions for the precompactness can obviously be relaxed.

THEOREM 10.1. *Suppose that for every compact set* $A \subset R^n$ *there exist two locally* $L_1$*-functions* $M_A(s)$ *and* $K_A(s)$ *such that if* $x, y$ *in* $A$ *and* $s \in R$:

(i) $|f(x, s)| \leqq M_A(s)$,

(ii) $|f(x, s) - f(y, s)| \leqq K_A(s)$,

*and such that $M_A$ and $K_A$ satisfy:*

(iii) *the family of $L_1$-functions $m_s(\tau):[0, 1]\to R$ for $s\in R$, given by $m_s(\tau)= M_A(s+\tau)$ is weakly precompact (uniformly integrable) in $L_1$,*

(iv) *the family of $L_1$-functions $k_s(\tau):[0, 1]\to R$ for $s\in R$, given by $k_s(\tau)= K_A(s+\tau)$ is bounded in $L_1$.*

*Then (*) is positively precompact, i.e., whenever $t_i\to\infty$ a subsequence of $f^{t_i}$ converges to an (ordinary) function $g=g(x, s)$.*

The proof of this theorem is the context of [3, Theorem 4.1]. It generalizes a previous result by Wakeman [41] where $M_A$ and $K_A$ are assumed to be constants. It seems that the result in Theorem 10.1 can be further generalized by replacing the Lipschitz condition (ii) with an appropriate equicontinuity condition.

**11. Ordinary equations are not enough.** We return now to the idea stated at the end of § 1. First we want to explain why the limiting behavior of solutions of

$$(*)\qquad \dot{x}=f(x, s)$$

might not be described by ordinary differential equations. However, if we allow other types of equations we get analogous results and cover a wider area and behavior.

A solution of (*) is an absolutely continuous function, in particular differentiable a.e. However, a sequence of differentiable, even $C^\infty$, functions can converge uniformly to a nowhere differentiable function. In the analysis of § 2 we pointed out the importance of the following situation. The translates $\mathbf{x}^{t_j}$ (recall that $\mathbf{x}^t(s)=\mathbf{x}(t+s)$) of the solution $\mathbf{x}$, converge to $\mathbf{y}$; the translates $f^{t_j}$ converge to $g$, then $\mathbf{y}$ is a solution of $\dot{x}=g(x, s)$. What if $\mathbf{x}^{t_j}$ converge to $\mathbf{y}$ but $\mathbf{y}$ is not differentiable a.e.? There is no ordinary differentiable equation $\dot{x}=g(x, s)$ which has $\mathbf{y}$ as a solution. Consequently, $f^{t_j}$ cannot converge and our theory fails. But there is still hope. The function $\mathbf{y}$ might be a solution of an equation which is not an ordinary differential equation. If we could associate a meaning to the convergence of $f^{t_j}$, or of $\dot{x}=f^{t_j}(x, s)$, to this equation, or alternatively allow limiting equations which are not ordinary differential equations, we might still retain the previous structure.

An example of an equation that might serve as a limiting equation for the o.d.e. (*) is

$$(11.1)\qquad x(\tau)=x(a)+\int_a^\tau g(x(s), s)\, d\eta(s),$$

where $\eta$ is a continuous measure on the real line. Clearly if we approximate $d\eta$ by $\theta(s)\, ds$, where $ds$ is the Lebesgue measure, then the o.d.e. (in its integral form)

$$(11.2)\qquad x(\tau)=x(a)+\int_a^\tau g(x(s), s)\theta(s)\, ds$$

is "close" to (11.1). (For a particular example, see [5, § 10].)

The mathematical problem in general is to embed the translates $\dot{x}=f^t(x, s)$ of (*) in a space of (unordinary) equations and to associate a convergence structure with this space such that certain properties hold. For the purposes of our paper, especially invariance, we shall be interested in the analogue of the continuous dependence result, Theorem 6.1.

In the next three sections we systematically treat some aspects of the theory. First (§ 12) we are looking for a general form of a limiting equation for (*), and identify the appropriate convergence for these equations. We also note a classification of these limiting equations. Then (§ 13) we state the invariance properties obtained for the enlarged class. In § 14 we give the conditions for positive precompactness, where unordinary equations are allowed. Proofs will not be given, since the complete theory appears in [4], [5].

## 12. Ordinary integral-like operator equations, definition, convergence and classification.

DEFINITION 12.1. An *ordinary integral-like operator H* is a mapping which associates with each $R^n$-valued continuous function $u$, and $a$ in the domain of $u$ a continuous function $H_a u$ so that:

(i) $H_a : C[a, b] \rightarrow C[a, b]$ is continuous for each interval $[a, b]$.

(ii) $H_a u(t) = H_a u(s) + H_s u(t)$ for all $s$, $t$ and $a$ in the domain of $u$.

We still assume that the ordinary equation

$$(*) \qquad \dot{x} = f(x, s)$$

satisfies Assumption (A). With the right-hand side $f$ we associate the ordinary integral-like operator $H$ defined by

$$H_a u(\tau) = \int_a^\tau f(u(s), s)\, ds.$$

Obviously, Assumption (A) implies that all the functions in the range of this $H_a$ are equicontinuous. This property will be transferred to the limiting equations. Therefore, we have:

DEFINITION 12.2. We say that the ordinary integral-like operator $H$ is *consistent with Assumption* (A) if whenever $u : [a, b] \rightarrow K \subset R^n$ is continuous the inequality $|H_a u(b)| \leqq \mu(b-a)$ holds.

DEFINITION 12.3. With the ordinary integral-like operator $H$ an equation

$$(**) \qquad u = Hu$$

is associated. A function $u(s)$ is a solution of (**) if whenever $s$ and $a$ are in its domain

$$u(s) = u(a) + H_a u(s).$$

The initial value problem $u = Hu$, $u(a) = x_0$ can be equivalently written as $u(s) = x_0 + H_a u(s)$. (Notice that $u = Hu$ is only a symbol, $Hu$ is not defined.)

Our candidates for limiting equations for (*) are equations (**) where $H$ is consistent with Assumption (A).

The reader might notice that the construction of our general form of the limiting equation is essentially a completion process of $\mathscr{G}$ (see § 4) with respect to the convergence in Definition 4.2. This immediately suggests a convergence for the operator equations.

DEFINITION 12.4. The sequence $f^{t_k}$ of translates, when $t_k \to \infty$, converges to the operator $H$ if whenever $u_k: [a, b] \to R^n$ is a sequence of continuous functions which converges uniformly to $u$, then $\int_a^b f^{t_k}(u_k(s), s)\, ds$ converges to $H_a u(b)$.

This convergence can be easily generalized to a convergence on the collection of integral-like operators which are consistent with Assumption (A) (see [5; Definition 5.1]). Also, we have the analogue of the continuous dependence result, Theorem 6.1, for this class of ordinary integral-like operator equations. We state it here for the case of translates converging to an operator.

THEOREM 12.5. *Let $y_k \to y_0$ in $R^n$, and $f^{t_k} \to H$, where $t_k \to \infty$. For each $k$ let $\boldsymbol{x}_k = x_k(s)$ be a maximally defined solution of $\dot{x} = f^{t_k}(x, s)$, $x(0) = y_k$. Then a subsequence $\boldsymbol{x}_m$ exists which converges to a maximally defined solution $u$ of $u(s) = y_0 + H_a u(s)$, and the convergence is uniform on compact intervals of the domain of $u$.*

(Cf. [5, §§ 4, 5].)

We want to state a completeness result. It says that if for $t_k \to \infty$ it is true that whenever $u_k: [a, b] \to R^n$ converge uniformly the sequence of vectors $\int_a^b f^{t_k}(u_k(s), s)\, ds$ converges, then the sequence $f^{t_k}$ converges to a certain ordinary integral-like operator equation which is consistent with Assumption (A) (see [5, Proposition 6.6]).

An interesting and important problem is to classify the ordinary integral-like operator equations that arise as limiting equations. It might be helpful to know that the limiting equations are of a certain particular type. One class that has been investigated is the one of Kurzweil equations. Kurzweil [18] developed a generalization of ordinary differential equations; he used it for various purposes including the explanation of certain continuous dependence phenomena (see [18], [19]). A self-contained study of Kurzweil equations and their relation to the subject of the present paper is given in [4]. The compactness result in [4] says that if (iii) in Theorem 10.1 above is relaxed to

(iii) *the family of $L_1$-functions $m_s(\tau): [0, 1] \to R$ has the property that $M_s(t) = \int_0^t m_s(\tau)\, d\tau$ are equicontinuous,*

then every limiting equation is a Kurzweil equation (but not necessarily an ordinary differential equation).

Another result in the direction of classification is that if (*) has a unique limiting equation, this limiting equation is actually an autonomous ordinary differential equation (see [5, Theorem 9.1 and Remark 9.2]). This result uses heavily Assumption (A) and it is not true without the assumption.

**13. Invariance properties and invariance principles with respect to the unordinary limiting equation.** We promised in § 11 to state the invariance properties for the $\omega$-limit sets where unordinary limiting equations are allowed. The reason why we will not keep this promise is that all the results, Theorems 7.1 through 7.6, hold when the ordinary differential limiting equations are replaced by ordinary integral-like limiting equations, and the same proofs apply.

We do want to re-examine Examples 7.7 and 7.8. The equation is of the form

$$\dot{x} = f_1(x, y, s),$$
$$\dot{y} = f_2(x, y, s)y + f_3(x, y) \tag{13.1}$$

in the $(x, y) \in R^n \times R^m$ space. We assume that $f_3(x, y) \neq 0$ if $x \neq 0$.

We claim that the conclusions of Examples 7.7 and 7.8 are true even if ordinary integral-like operator equations are the limiting equations. This will certainly relax the conditions on $f$—both the existence of a limiting equation in Example 7.8 and the positive precompactness in Example 7.7 are more easily fulfilled.

To justify the claim we note that the use of the limiting equations was to show that the only invariant set in $E = \{(x, y) : y = 0\}$ is the origin $\{0\}$. Now, it is easy to see, that if $H$ is a limiting operator of (13.1), and $(u(s), 0)$ is a function, then the $y$-coordinates of $H_a(u, 0)$ are given by $\int_a^\tau f_3(u(s), 0)\, ds$. This is an immediate consequence of the convergence in Definition 12.4. The assumptions of $f_3$ imply that the only invariant subset of $E$ is $\{0\}$.

Note that we did not have to compute the limiting integral-like operators. Their existence together with the structure of the equation yield the information.

**14. Positive precompactness in the wide sense.** Recall that

$$(*) \qquad \dot{x} = f(x, s)$$

is positively precompact if for every sequence $t_k \to \infty$ a subsequence of $f^{t_k}$ converges. If we allow a wider class of limiting equations we obviously get more relaxed conditions for the precompactness (cf. Theorem 10.1).

THEOREM 14.1. *Suppose that for each fixed s and a fixed compact set $K \subset R^n$ the function $f(\cdot, s)$ satisfies*

$$|f(x, s) - f(y, s)| \leqq \nu_K(|x - y|, s),$$

*where $\nu_K(\cdot, s)$ is nondecreasing, continuous at* 0 *and $\nu_K(0, s) = 0$. Also suppose that the moduli $\nu_K(r, s)$ are locally integrable in s, and*

$$\int_s^{s+1} \nu_K(r, \tau)\, d\tau \leqq N_K(r)$$

*for every s, and $N_K(r) \to 0$ as $r \to 0$. Then* (*) *is positively precompact.*

The proof is given in [5, § 8]. There is still much room for improvement.

**15. On the convergence.** We shall discuss the convergence which we are using and shall give some answers to the questions raised in § 4.

It will be very useful to know that the convergence is generated by a metric or a topology. It will enable us to use continuity concepts from point-set topology. Unfortunately, there is no metric or topology on $\mathcal{G}$ which generates the convergence. We shall sketch an argument which is a paraphrase of [1, Appendix B].

For $m = 1, 2, \cdots$ let $h_m(t)$ be a piecewise linear function from $R$ to $[0, 1]$, with "pieces" connecting the points $(2k/m, 0)$ and $((2k+1)/m, 1)$ in $R \times [0, 1]$. For $n, m = 1, 2, \cdots$ let $\phi_{n,m}$ be the function $\phi_{n,m}(s) = (1/n)h_m(s)$. Let $g_{n,m}(x, s)$ be defined on $R \times R$ by: $g_{n,m}(x, s) = 1$ if $x = \phi_{n,m}(s)$; $g_{n,m}(x, s) = 1/m$ if $|x - \phi_{n,m}(s)| \geqq 1/m$; and on the rest of $R \times R$, $g_{n,m}$ is a continuous extension into $[0, 1]$. As is done in [1, Appendix B] one shows that $g_n \equiv 1/n$ is the limit of $g_{n,m}$ as $m \to \infty$. So $g_n$ is in the closure of $\{g_{n,m} : n, m = 1, 2, \cdots\}$. But $g \equiv 0$, which is in the closure of the closure, is not in the closure itself.

Something still can be said. There is a topology whose converging *sequences* coincide with the converging sequences of $\mathcal{G}$. This topology is the compact-open topology, where the elements of $\mathcal{G}$ are regarded as operators from $C[a, b]$ into $C[a, b]$ as was done in § 12. So sequential continuity with respect to the compact-open topology is the continuity with respect to the convergence on $\mathcal{G}$.

Another direction is to find subclasses of $\mathcal{G}$ on which the convergence is given by a metric. There is much to be done in this direction. A partial result is indirectly given in [3], [4]. For instance, for functions that satisfy conditions (i)–(iv) in Theorem 10.1 (with the same $M_A$ and $K_A$) the convergence of $g_k$ to $g_0$ in $\mathcal{G}$ is equivalent to

$$\int_a^b g_k(x, s)\, ds \to \int_a^b g_0(x, s)\, ds \tag{15.1}$$

for every interval $[a, b]$ and every vector $x \in R^n$. The convergence given in (15.1) is then generated by a metric (see [3] and also [41]).

With regard to the second question, formula (15.1) gives a better representation, but for a particular subclass of $\mathcal{G}$.

Several types of convergences were used in the literature in connection with problems similar to our research. Miller [27] and Sell [36], [39], have used the uniform convergence on compact intervals of $g_k(x, s)$ to $g_0(x, s)$ in the construction of the limiting equations. An integral criterion was used by Strauss and Yorke [40], and by Miller and Sell [29], [30], in the context of integral equations, and it was generalized by Neustadt [32]. Rouche [35] has also investigated limiting equations generated by an integral convergence, which is similar to the $L_1$-convergence. We require a sort of joint continuity with respect to a weak $L_1$-convergence. The particular form of (15.1), which applies to equicontinuous functions, appeared already in Gikhman [12] in connection with continuous dependence, and since then was relaxed by several authors.

The last question is: Can the convergence be relaxed? We, of course, want to maintain the conclusions. In our context an anwer depends on finding necessary conditions for the continuous dependence of solutions on initial data and on the right-hand side of the equation (see § 6). The author has done some work on this problem. In short, the convergence in Definition 4.2 is not a necessary condition for the continuous dependence as the continuous dependence is stated. But it is a necessary condition if we demand that solutions of

$$x(t) = z(t) + \int_0^t g(x(s), s)\, ds$$

will depend continuously on $g$ and on the continuous *function* $z(t)$. So by requiring a little bit more from the type of continuous dependence, our convergence becomes a necessary condition (cf. [1], [2] and [5, § 5]).

**16. Some remarks on the literature and related topics.** Our paper deals mainly with limiting equations and their relation to invariance properties and stability theory. The limiting equations form a useful tool in various other fields of interest. A particular and important one is related to the applications of classical topological dynamics to the study of nonautonomous differential and integral equations. We are not going to compete with Sell's monograph [37] and the surveys by Sell [38] and recently by Miller and Sell [31].

The particular case of asymptotically autonomous systems was already discussed by Markus [26], and in the same spirit as in our present paper, namely, deriving properties of the original equation from properties of the limiting equation. Limiting analysis of equations which are not asymptotically autonomous was done by Miller [27], [28] for almost periodic systems. Sell [36] gave the foundations of the translate techniques and the relation to classical topological dynamics. Miller and Sell [30] generalized it to Volterra integral equations. A particular case of o.d.e.'s without uniqueness was treated by Sell [39].

The invariance principle and its importance to stability was discovered by LaSalle [20]–[23]; except in [21] (where periodic equations are discussed), the autonomous case was developed. Related work and applications can be found in Onuchic [33] and Peng [34]. The invariance for general nonautonomous systems was done by Dafermos [9], [10] in the context of flows and dynamical systems of solutions (it was related in [11] to the dynamical systems of Sell [36]). Wakeman [41] has established an invariance principle in the form of our results in § 7. See also Rouche [35] and LaSalle's survey [24]. The conditions are more relaxed in [3]. Unordinary limiting equations were introduced in [4] and [5].

**Acknowledgment.** I wish to express my thanks to Professor LaSalle for his advice and encouragement while carrying out this research and preparing these notes.

# References: Appendix A


[1] Z. ARTSTEIN, *Continuous dependence of solutions of Volterra integral equations*, SIAM J. Math. Anal., 6 (1975), pp. 446–456.

[2] ———, *Continuous dependence of fixed points of condensing maps*, Dynamical Systems. An International Symposium. Academic Press, New York, 1976.

[3] ———, *Topological dynamics of an ordinary differential equation*, J. Diff. Eqs., to appear.

[4] ———, *Topological dynamics of ordinary differential equations and Kurzweil equations*, Ibid., to appear.

[5] ———, *The limiting equations of nonautonomous ordinary differential equations*, to appear.

[6] ———, *Limiting equations and Liapunov functions*, forthcoming.

[7] Z. ARTSTEIN AND E. F. INFANTE, *On the asymptotic stability of oscillators with unbounded damping*, Quart. Appl. Math., to appear.

[8] T. A. BURTON, *An extension of Liapunov's second method*, J. Math. Anal. Appl., 28 (1969), pp. 545–552; a correction: Ibid, 32 (1970), pp. 689–691.

[9] C. M. DAFERMOS, *An invariance principle for compact processes*, J. Diff. Eqs., 9 (1971), pp. 239–252.

[10] ———, *Uniform processes and semicontinuous Liapunov functionals*, Ibid., 11 (1972), pp. 401–415.

[11] ———, *Semiflows associated with compact and uniform processes*, Math. Systems Theory, 8 (1974), pp. 142–149.

[12] I. I. GIKHMAN, *On a theorem of N. N. Bogolyubov*, Ukrain. Math. J., 4 (2) (1952), pp. 215–218.

[13] J. R. HADDOCK, *On Liapunov functions for nonautonomous systems*, J. Math. Anal. Appl., 47 (1974), pp. 599–603.

[14] ———, *Stability theory for nonautonomous systems*, Dynamical Systems. An International Symposium. Academic Press, New York, 1976.

[15] J. K. HALE, *Ordinary Differential Equations*, Wiley–Interscience, New York, 1969.

[16] P. HARTMAN, *Ordinary Differential Equations*, John Wiley, New York, 1964.

[17] C. KURATOWSKI, *Topology I*, Academic Press, New York, 1966.

[18] J. KURZWEIL, *Generalized ordinary differential equations and continuous dependence on a parameter*, Czech. Mat. J., 7 (82) (1957), pp. 418–449; an addition: Ibid., 9 (84) (1959), pp. 564–573.

[19] ———, *Problems which lead to a generalization of the concept of an ordinary differential equation*, Differential Equations and Their Applications, Proc. Conference Prague, September 1962, Academic Press, 1963, pp. 65–76.

[20] J. P. LASALLE, *The extent of asymptotic stability*, Proc. Nat. Acad. Sci. U.S.A., 46 (1960), pp. 363–365.

[21] ———, *Asymptotic stability criteria*, Proc. Symp. Appl. Math. Hydrodynamic Instability, vol. 13, Amer. Math. Soc., Providence, R. I., 1962, pp. 299–307.

[22] ———, *An invariance principle in the theory of stability*, Differential Equations and Dynamical Systems, Proc. Internat. Symp., Puerto Rico, Academic Press, New York, 1967, pp. 277–286.

[23] ———, *Stability theory for ordinary differential equations*, J. Diff. Eqs., 4 (1968), pp. 57–65.

[24] ———, *Stability theory and invariance principles*, Dynamical Systems. An International Symposium. Academic Press, New York, 1976.

[25] J. J. LEVIN, *On the global asymptotic behavior of nonlinear systems of differential equations*, Arch. Rational Mech. Anal., 6 (1960), pp. 65–74.

[26] L. MARKUS, *Asymptotically autonomous differential systems*, Contribution to Nonlinear Oscillations, vol. III, S. Lefschetz, ed., Princeton University Press, Princeton, N.J., 1956, pp. 17–29.

[27] R. K. MILLER, *Almost periodic differential equations as dynamical systems with applications to the existence of a.p. solutions*, J. Diff. Eqs., 1 (1965), pp. 337–345.

[28] ———, *Asymptotic behavior of solutions of nonlinear differential equations*, Trans. Amer. Math. Soc., 115 (1965), pp. 400–416.

[29] R. K. MILLER AND G. R. SELL, *Existence, uniqueness and continuity of solutions of integral equations*, Ann. Mat. Pura. Appl., 80 (1968), pp. 135–152; 87 (1970), pp. 281–286.

[30] ———, *Volterra integral equations and topological dynamics*, Mem. Amer. Math. Soc., no. 102, 1970.

[31] ———, *Topological dynamics and its relation to integral equations and nonautonomous systems*, Dynamical Systems. An International Symposium. Academic Press, New York, 1976.

[32] L. W. NEUSTADT, *On the solutions of certain integral-like operator equations. Existence, uniqueness and dependence theorems*, Arch. Rational Mech. Anal., 38 (1970), pp. 131–160.

[33] N. ONUCHIC, *Invariance properties in the theory of ordinary differential equations with applications to stability problems*, SIAM J. Control, 9 (1971), pp. 97–104.

[34] T. K. C. PENG, *Invariance and stability for bounded uncertain systems*, SIAM J. Control, 10 (1972), pp. 679–690.

[35] N. ROUCHE, *The invariance principle applied to non-compact limit sets*, Boll. Un. Mat. Ital., to appear.

[36] G. R. SELL, *Nonautonomous differential equations and topological dynamics. I, II*, Trans. Amer. Math. Soc., 127 (1967), pp. 241–283.

[37] ———, *Lectures on Topological Dynamics and Differential Equations*, Van Nostrand–Reinhold, London, 1971.

[38] ———, *Topological dynamics techniques for differential and integral equations*, Ordinary Differential Equations, Proc. 1971 NRL-MRC Conf., L. Weiss, ed., Academic Press, New York, 1972, pp. 287–304.

[39] ———, *Differential equations without uniqueness and classical topological dynamics*, J. Diff. Eqs., 14 (1973), pp. 42–56.

[40] A. STRAUSS AND J. A. YORKE, *On asymptotically autonomous differential equations*, Math. Systems Theory, 1 (1967), pp. 175–182.

[41] D. R. WAKEMAN, *An application of topological dynamics to obtain a new invariance property for nonautonomous ordinary differential equations*, J. Diff. Eqs., 17 (1975), pp. 259–295.

[42] T. YOSHIZAWA, *Asymptotic behavior of solutions of a system of differential equations*, Contrib. Diff. Eqs., I (1963), pp. 371–388.

[43] ———, *Stability Theory by Liapunov's Second Method*, Publication no. 9, Math. Soc. of Japan, Tokyo, 1966.